WILD MALAYSIA

WILD MALAYSIA

The wildlife and scenery of Peninsular Malaysia, Sarawak and Sabah

---

Photographs by **GERALD CUBITT**
Text by **JUNAIDI PAYNE**

Produced in association with the World Wide Fund for Nature Malaysia

First published in the United Kingdom in 1990 by
New Holland (Publishers) Ltd

London • Cape Town • Sydney • Singapore

24 Nutford Place   80 McKenzie Street   3/2 Aquatic Drive
London W1H 6DQ   Cape Town 8001   Frenchs Forest, NSW 2086
United Kingdom   South Africa   Australia

Reprinted in 1992, 1994 and 1996

ISBN: 1 85368 093 1

Project Manager: Charlotte Parry-Crooke
Editor: Ann Hill
Editorial Assistants: Vicky Cooper, Tracey Williams
Designer: Behram Kapadia
Cartography: Julian Baker, Maltings Partnership
Index: P.E. Barber

Typeset by AKM Associates (UK) Ltd, Southall, London
Reproduction by Chroma Graphics (Overseas) Pte Ltd, Singapore
Printed and bound in Singapore by Kyodo Printing Co (Singapore) Pte Ltd

CONTENTS

FOCUS ON
PENINSULAR MALAYSIA

FOCUS ON SARAWAK

FOCUS ON SABAH

Photographic Acknowledgements

The publishers and photographer extend their thanks to the following people who kindly loaned their photographs for inclusion in this book. All the photographs in the book, with the exception of those listed below, were taken by Gerald Cubitt.

C.M. Francis: page 37 (above); page 145 (above right).

Ron Holland (Borneo Divers): page 39; page 200 (all five subjects); page 201 (above and left).

Tony Lamb: page 144 (all four subjects); page 145 (above left); page 150 (below left); page 172 (below left); page 186 (below left); page 195 (above right); page 196 (top and bottom).

Frank Lambert: page 78 (below left and below right); page 79 (all three subjects).

MAS, Malaysia: page 150 (below right).

Junaidi Payne: page 21; page 25 (right); page 26 (both subjects); page 46 (above); page 189 (above).

Slim Sreedharan: page 75 (left); page 195 (above left).

Ho Cheng Tuck: page 194 (below).

Illustrations appearing in the preliminary pages are as follows:

HALF-TITLE: Rajang River, Sarawak, at sunrise.

FRONTISPIECE: Male Orang-utan at the Sepilok Forest Reserve, Sabah.

PAGE 5: Sunset in the Langkawi Islands.

PAGE 6: Female Proboscis Monkey, Kinabatangan River.

PAGE 7: Mount Kinabalu at sunset.

PAGES 12–13: Darvel Bay, Sabah.

Acknowledgements

Interest in this project has been forthcoming from numerous sources around the world. The publishers would especially like to thank the sponsors, contributors and consultants for their involvement. The author, photographer and publishers would like to express their gratitude to the following for their generous and valuable assistance during the preparation of the book:

World Wide Fund for Nature (WWF)

World Wide Fund for Nature (WWF) Malaysia

Peter Jackson

Malaysia Airlines

Petronas

Tourist Development Corporation, Ministry of Culture and Tourism, Malaysia

Peninsular Malaysia

Mohd. Khan bin Momin Khan, Director General, Department of Wildlife and National Parks · The staff of Taman Negara National Park
Dr Salleh Mohd. Nor, Director General, Forest Research Institute, Malaysia
Dr Tho Yow Pong, Forest Research Institute, Malaysia
Malayan Nature Society · Dr Kiew Bong Heang · Victor and Micky Smets
Susan Abraham, WWF Malaysia · David Teng · Heah Hock Heang
Rosemarie Wee, Public Relations Manager, Shangri-La Hotel, Kuala Lumpur
Shangri-La Hotels: Kuala Lumpur and Penang (Golden Sands)
Merlin Hotel Group · Tioman Island Resort

Sarawak

Philip Ngau Jalong
National Parks and Wildlife Office, Sarawak Forestry Department
David Labang
Dr Elizabeth Bennett, New York Zoological Society/WWF Malaysia
Holiday Inn, Kuching

Sabah

Ministry of Tourism and Environmental Development
Sabah Wildlife Department · Francis Liew, Sabah Parks
Sam Mannan, Sabah Forestry Department
Dr Clive Marsh, Principal Forestry Officer (Conservation), Sabah
Foundation · Tony and Anthea Lamb · Karen Phillipps
Dr Rob Stuebing, Universiti Kebangsaan Malaysia (Sabah campus)
Hyatt Kinabalu International

Singapore

Peter Ng, Singapore University · Bernard Harrison · Roy Sirimane

Great Britain

The Trustees and Staff of the Natural History Museum, London
Dr John Dransfield, Royal Botanic Gardens, Kew
Dr Roy Watling, Royal Botanic Garden, Edinburgh
Dr George Argent, Royal Botanic Garden, Edinburgh
Rosemary Smith, Royal Botanic Garden, Edinburgh
Dr Alwyne Wheeler, Epping Forest Conservation Centre
Martin Woodcock

Special Acknowledgement

The publishers, author and photographer would like to express special thanks to Ken Scriven, Executive Director of the World Wide Fund for Nature Malaysia, for his advice, encouragement and support during the preparation of this book.

PREFACE

Malaysia has changed enormously during the past thirty years and Malaysians are better off than before. Our country is often regarded as a model developing nation being able to strike a balance between conservation and development.

Malaysians are proud of their achievements in agriculture and industry. Perhaps we have been too modest to express an equally justifiable pride in our heritage. The hills, the forests and the flora and fauna of Malaysia give us much – from enjoyment of beauty to a reminder of God's creation, from clean water to new insights for scientific and technological progress.

This publication depicts both the beauty and the diversity of our country. I hope it will serve to demonstrate that Malaysia's remarkable natural heritage is indeed a part of our national heritage and that despite the rapid development taking place, areas of historical and environmental importance are being conserved and protected.

DR MAHATHIR BIN MOHAMAD
PRIME MINISTER
MALAYSIA

WWF

FOREWORD

Malaysia has been endowed with a natural heritage of great richness and splendour. The great tropical rainforests with their wealth of wildlife, both plant and animal, once covered the entire country.

Over the past decades the demands of development have taken their toll and although the visitor is still impressed by the lush greenness that is everywhere, he or she rarely has the opportunity to journey up the great rivers into the heart of the rainforest.

Those of us who are deeply concerned with conservation are, however, frequently privileged to make such journeys in the course of our endeavours to save, for all time, representative examples of our heritage. As this book portrays, the natural beauty and wonders of Malaysia continue to exist for those who have a spirit of adventure and who are prepared to travel to the more remote areas.

Malaysia has a variety of national parks, wildlife sanctuaries and other protected areas that are increasingly becoming recognized as being of importance to the conservation of the environment and as attractions for the tourist.

I am confident that *Wild Malaysia* will contribute to the expansion of knowledge about our natural heritage which those of us in WWF Malaysia are dedicated to preserving for the future benefit of all.

TAN SRI KHIR JOHARI
PRESIDENT
WWF MALAYSIA

INTRODUCTION

———◆———

Malaysia is a country with a rich diversity of peoples and cultures for which it is justly famous. Driving along the broad highways which link the international airports with busy urban centres, one is offered tantalizing glimpses of these people, their towns and villages, their multifarious activities. It is therefore sometimes difficult to remember that only a century ago this was a sparsely populated land, covered by a vast expanse of untouched tropical rainforest. Today much of the virgin forest has been cleared to make way for agriculture, new plantations and industrial developments. The export of hardwoods from the rainforest has for many years constituted a major source of revenue. Yet despite these inroads, Malaysia remains a land of spectacular scenery, with some unforgettable sights and sounds. The tropical islands, fringed by coral reefs and set in a clear blue sea, are as idyllic as anywhere in the world. Sunset over the South China Sea, as the colours across the water slowly change from brilliant blues, yellows and reds to softer, more subtle shades, evokes a sensation of utter tranquillity. Another memorable experience is to travel up one of the inland rivers where the forested banks remain undisturbed by any human activity. The beautiful Sungai Tahan in Taman Negara is quite untouched. Once numerous throughout Malaysia, sights such as this will survive only in a few protected areas, once the remaining forest reserves have been logged for timber. Finally, to stand in the rainforest at dawn looking up into the tall tree canopy and hearing the cries of the gibbons, the chattering of the squirrels, the ceaseless humming of the insects and the pooping call of the Helmeted Hornbill is perhaps the quintessential Malaysian experience.

The wildlife itself, in closer focus, is more difficult of access. In rainforests, many of the country's natural treasures lie hidden from our sight. Nature is intricate, complicated, often subdued, needing to be sought out not only through visual observation but also with a sensitivity towards its sounds, scents, textures and flavours. Discovering these natural beauties presents us with a challenge, but one that is richly rewarding.

Malaysia is a federation of thirteen states, eleven of which are situated on South-east Asia's southernmost peninsula and offshore islands, and known collectively as Peninsular Malaysia. The eleven states occupy an area of 131,235 square kilometres (or 50,670 square miles). The two other states of Sarawak (124,450 square kilometres or 48,050 square miles) and Sabah (76,115 square kilometres or 29,388 square miles) are located on the island of Borneo. These sections, often referred to as 'West Malaysia' and 'East Malaysia', are separated by over five hundred kilometres (310 miles) of the South China Sea at their nearest point. They possess close basic similarities of flora, fauna and of peoples. Malay has been the major language of the region for many generations. A combination of migrations by Moslem Malay-speaking peoples, of vigorous settlement and plantation agriculture by the British, and of migrations of Chinese and Indians, has transformed the face of Peninsular Malaysia since the end of the last century. Sarawak and Sabah have changed less, later and more slowly. The last official census showed nearly seventeen million people living in Malaysia, with about fourteen million concentrated in the Peninsula.

In the old days, the sea and rivers were the main routes of communication and trade, but their importance has since been almost totally supplanted by road and air travel. In Peninsular Malaysia the roads are extensive and excellent, but in Sarawak and Sabah they are limited and often rough, so that long-distance travel in these states may start as an adventure but end in exhaustion. The Peninsula road system serves mainly to link towns and villages by way of vast stretches of plantation and deforested land, while in Sarawak and Sabah most public roads pass through or close to the forest, or to land supporting traditional agriculture and life-styles. In Malaysia, Sarawak alone has several large river systems which are navigable far into the interior and which have public boat services (at least on the lower stretches). Most of Malaysia is well-served by regular aeroplane flights and, with careful planning, it is possible to see many of the wild parts of the country within the constraints of a modest budget and limited time.

OPPOSITE PAGE The essence of wild Malaysia – an island near Semporna, off Sabah's east coast.

Climate, Geography and Geology

Malaysia lies within the tropical zone, where it is hot throughout the year. It is this hot, wet, benign climate which has played a large part in moulding natural, wild Malaysia. Daytime temperatures in the lowlands average a bearable 30°C (86°F) and rarely exceed 32°C (90°F), but the beneficial cooling effect of moisture evaporating from the human body is countered by the high humidity. Temperatures are lower in the forest but it is here that humidity becomes downright uncomfortable. For every 100-metre (330-foot) increase in altitude, the air temperature drops by roughly 0.6°C (1°F). Freezing-point is reached only rarely, and at only one spot in Malaysia – on the top of Mount Kinabalu.

While some continental areas of the tropics are dry, Malaysia, surrounded on all sides by sea, is bathed in a flow of permanently warm, moist air that brings humidity and rain throughout the year. Two major wind systems influence the region's weather, causing some seasonality in rainfall. The north-east monsoon winds bring extra rain during the period from October to March, especially to Sarawak and the eastern parts of Peninsular Malaysia and Sabah. During this time of the year the sky is grey, the landscape drenched and the rivers swollen. The south-west monsoon has more force on the western sides of Peninsular Malaysia, Sarawak and Sabah, bringing rain from April to August. There is usually no dry season, and periods known locally as 'dry seasons' are merely times of low rainfall. The land never has time to become really thirsty, although the annual rainfall varies widely throughout the country and there may be appreciable differences between sites only a few kilometres apart. Interior Sarawak has the highest total: 350–500 centimetres (140–200 inches) a year. By contrast, the interior plains, sheltered by ranges of hills, typically receive less than 200 centimetres (80 inches) annually. Most crops tend to do best where the total rainfall is less than 250 centimetres (100 inches). It is largely the ability of oil palm and rubber trees to thrive in wetter conditions that has led to their successful large-scale cultivation throughout Malaysia.

An exceptionally long period with very little rain occurred in Sabah and East Kalimantan (the Indonesian part of Borneo) from late 1982 to mid-1983. During these months, massive forest fires raged on a scale never before seen in the tropics. The causes were readily apparent: the inhabitants had little experience of large-scale fires, and vast areas of forest, which had been commercially logged, were littered with dry, dead wood. It has been estimated that up to 10,000 square kilometres (over 3,850 square miles) of Sabah's forests were damaged by fire at that time. Analysis of old rainfall records has shown that periods of exceptionally low rainfall do occur in most localities in Malaysia once every few decades. In the past, however, the forests were untouched, and there was little dry dead wood lying around to spread any fires that did occur. On the other hand, even a moderate drought without any fire could have severe consequences on those plant species which require continuous moisture. Many wild plants in Malaysia – especially palms – can be found in only a few, scattered locations, and some botanists believe that this may be because past droughts have wiped them out.

Although monsoon winds influence Malaysia's rainfall patterns, strong winds are not a feature of the land, and typhoons

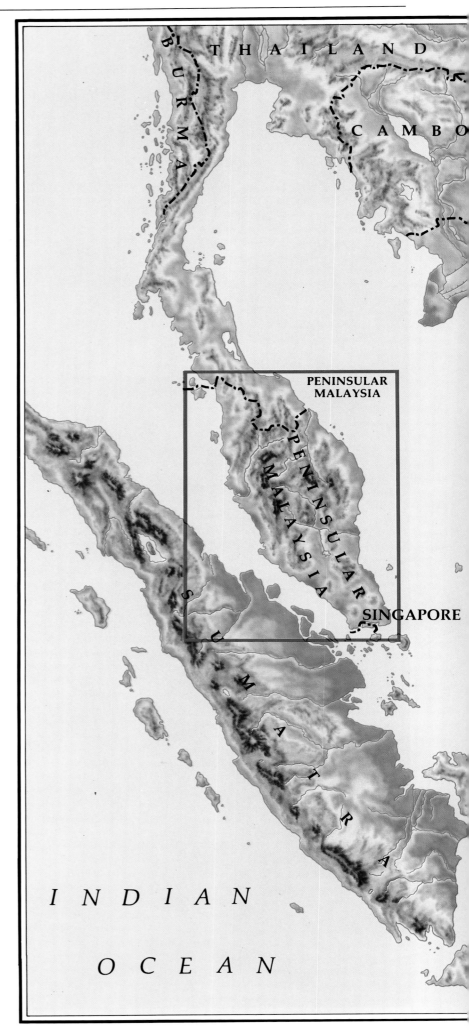

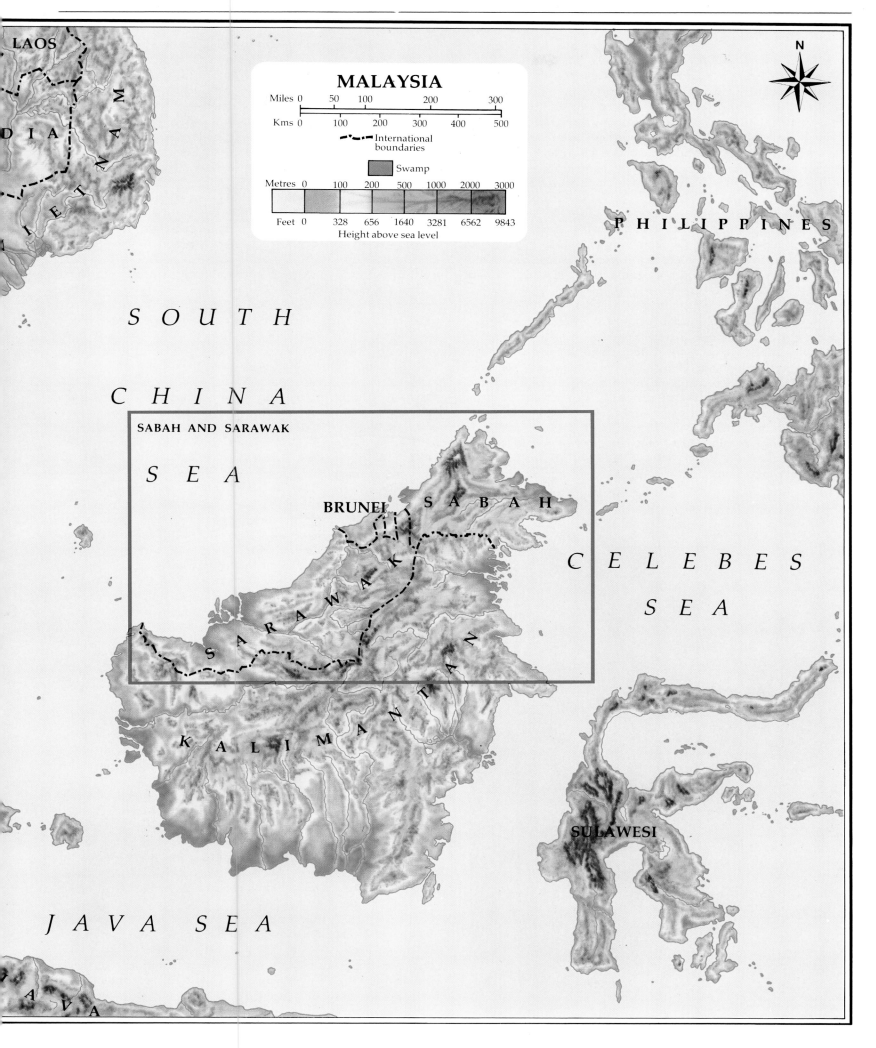

MALAYSIA

Miles 0   50   100   200   300

Kms 0   100   200   300   400   500

International boundaries

Swamp

Metres 0   100   200   500   1000   2000   3000

Feet 0   328   656   1640   3281   6562   9843

Height above sea level

LAOS

VIETNAM

DIA

PHILIPPINES

SOUTH

CHINA

SEA

SABAH AND SARAWAK

SEA

BRUNEI   SABAH

CELEBES

SARAWAK

SEA

SARAWAK

KALIMANTAN

SULAWESI

JAVA SEA

JAVA

miss the country by several hundreds of kilometres. Viewed from the populated and cultivated lowlands, the forest-clad hills and mountains sleep in a distant blue haze, never showing anger, for the last spate of volcanic activity occurred more than 20,000 years ago. These are conditions where plants and insects and other small organisms reproduce and grow with vigour, having evolved over the millennia into countless specialized forms. Human beings have a tougher time in this environment, for it is difficult to make any order out of the complicated jumble of nature, difficult to live in harmony with it, difficult even to keep cool.

Had technology reached its present levels 15,000 years ago, it would have been possible to build a road from Peninsular Malaysia, via Sumatra and Kalimantan in Indonesia, to Sarawak and Sabah. At that time, temperatures were a few degrees lower than now, sufficient to lock up vast amounts of the world's water as ice, both in the temperate climatic regions of the earth and on the mountain tops of the tropics. The sea level was at least fifty metres (160 feet) lower than it is now and instead of being an archipelago, the islands of Borneo, Sumatra and Java, as well as Peninsular Malaysia, were part of a single land-mass which scientists call Sundaland. The Philippine islands and Sulawesi were even then isolated, being separated by very deep submarine trenches. During alternating ice ages and warm periods, Sundaland was repeatedly formed and broken up by changes in sea level. Plants and animals were able to expand their ranges of distribution quite easily when Sundaland lay exposed above the sea. This accounts for the close similarities of flora and fauna in the region, and possibly also for some of the strange differences.

Limestone outcrops at Niah, Sarawak.

Many people have puzzled as to why tigers occur in Peninsular Malaysia, Sumatra and Java but not in Borneo. Possibly the species expanded its range southwards only during the last ice age, after rising sea levels had cut off Borneo. In contrast, that strange little primate called the Tarsier, which now occurs only in southern Sumatra, Borneo, Sulawesi and the southern Philippines but not in Peninsular Malaysia, appears to be a remnant of a much older fauna, dating back millions of years, when the character of the region was very different from what it is today.

Peninsular Malaysia and Sabah share some physical similarities: a cultivated western coastal plain followed, further inland, by a rugged forested mountain range, which breaks up into numerous separate hill ranges, with meandering valleys and plains partly forested and partly cultivated on the eastern side. Both areas have a mixed geology. Sarawak has extensive lowland coastal plains of peat, but most of the state consists of rugged hill ranges, drained by a few massive river systems. The majority of rocks can be broadly classified into two main types: those which have been formed through the action of extreme heat and pressure, generally below the earth's surface (the so-called igneous and metamorphic rocks), and those which have been formed by rock particles accumulating on the surface of the earth, usually under the sea, over millions of years, the sedimentary rocks such as sandstones and mudstones. Movements of the earth over further millions of years have broken and jumbled the sediments, and exposed them at a variety of angles. On the earth's surface, different layers of sediments erode at different rates, forming the rugged terrain of the hill ranges, especially in Sarawak and Sabah. It is often possible to spot the difference between the hill ranges when flying over them. Hills of igneous or metamorphic rock are more rounded, with a rather even forest canopy, while those of sedimentary rocks are rugged, with an uneven canopy. Dramatic geological features crop up here and there. In Sabah, an enormous mass of granite was pushed up through the sedimentary rocks of the Crocker Range, starting about two million years ago. This granite pluton, as it is called by geologists, is the top part of Mount Kinabalu, now 4,101 metres (13,455 feet) above sea level at the highest peak and rising still. Limestone outcrops, some remnants of older geological periods, poke through the newer sediments in widely scattered localities. Many of the outcrops contain caves. They were home to people thousands of years ago, but are now inhabited only by bats, swiftlets and various other small creatures. Two of the world's most spectacular natural limestone features occur in the Gunung Mulu National Park in northern Sarawak: the largest-known cave system, and a bizarre array of tall limestone pinnacles which protrude eerily through the mist above the forest canopy.

Malaysia has deposits of some commercially valuable minerals and metals. Petroleum, a major source of government revenue, comes from deposits under the South China Sea, off the east coast of Peninsular Malaysia and the west coast of Sarawak and Sabah. The Peninsula has tin deposits, a major reason for its rapid early development. Sarawak and Sabah have coal, gold, copper and other metals, but in most areas extraction costs make exploitation uneconomic.

In many parts of the world, dry climate and eroded landscapes give rise to spectacular scenery, but scenery which is barren and dull when seen close to. The natural treasures of Malaysia, clothed as they are in lush forests and plantations, gardens, ricefields and townships, remain largely hidden from our eyes. But approach closer and you will discover a land whose physical aspect is as varied and as beautiful as any in the world.

Plant Life

Wild Malaysia is a land of conditions almost perfect for luxuriant plant growth. It is reckoned that Peninsular Malaysia has about 8,500 species of vascular plants (by which is meant flowering plants, plus ferns and their allies), while Borneo may have 11,000, of which most occur in Sarawak and Sabah. The fastest growth rate for a tree recorded anywhere in the world comes from Sabah, where *Albizzia falcataria* – actually not a native – has achieved a height of 10.74 metres (over 35 feet) in only thirteen months. The largest leaf ever recorded was of an aroid plant, *Alocasia macrorrhiza*, also a non-native, found near Tawau in Sabah. It was more than three metres long by 1.9 metres wide (10 feet by 6 feet). Smaller aroids abound in Malaysian forests, even in the mountain ranges: they are non-woody plants, often growing in damp, shady places, with rather fleshy, heart-shaped leaves. Curiously, they do not comprise a well-known group of plants, even among botanists.

Trees, lianas, epiphytes (plants which grow on other plants), palms and herbaceous plants all grow together to form forests. Every situation provides a niche for seeds and spores to germinate and for plants of one kind or another to grow continuously through the year. In the lowlands, trees are draped and effectively joined together by woody climbing plants called lianas, and some are strangled to death in very slow motion, eventually to be totally enclosed and replaced by giant fig plants. Slow-moving water is clogged by the water-hyacinth (an invader from South America). In the mountains, where moisture condensing from warm air brought up from the lowlands provides an ever-wet environment, mosses and orchids festoon the branches and trunks of trees. Any land cleared of vegetation is quickly carpeted by shrubs, herbs, grasses, sedges or scrambling plants, according partly to soil conditions and partly to which seeds arrived first. Anything which looks like grassland, bush or scrub should therefore be viewed with suspicion. Only the cloud-cloaked mountain tops and stretches of moving water have escaped invasion by vigorous and intrusive plant life.

Malaysia's forests mean different things to different people and have been called by many names. For those who have suffered here in times of war, they have been a 'green hell'. Those for whom the forest is home give it no general name but only names for specific areas. Tropical rainforest or tropical evergreen moist forest are now acceptable general descriptions. The form and composition of Malaysian rainforests varies greatly, however, mainly according to characteristics of the soil, slope and altitude. Broadly, most forests in Malaysia can be divided into five groups: mangrove and nipa in the coastal regions, freshwater swamp forests, dipterocarp forests, heath forests and montane forests. Human influence is important in many areas, and in

Plant life of the understorey in lowland dipterocarp forest, Sabah.

order to specify that a forest has not been substantially tampered with, botanists and foresters use the extra descriptive terms of primary, tall or virgin forest. Forests from which trees have been removed for commercial sale are known as logged forests, and where only certain trees are taken – generally the largest – the term selective logging is used. This is the usual form of logging in most Malaysian forests. If all trees are cut in a forest, the term clear-felling is used. Unlike many types of forest in temperate climates, tropical rainforests cannot be managed for timber production by clear-felling. Instead, where the orginal forest has been clear-felled and not deliberately replaced with a crop plant (for example, oil palm or cocoa), a different kind of vegetation grows up, consisting of a few species of fast-growing trees which never achieve great size. This is called secondary forest.

The Malaysian rainforest has sometimes been described as 'the oldest forest in the world'. This is rather misleading. Certainly, the forest system and its enormous wealth of species have been evolving for many tens of millions of years. But the extent and distribution of different types of forest must have varied greatly during this long period, according to climate and changing sea levels and, on a larger time-scale, with geological processes. For example, the close similarities between the plants on the upper levels of Mount Kinabalu and mountain systems elsewhere can only be explained by assuming that what are now strictly mountain flora were once distributed widely throughout Asia.

The Coastal Fringe

Mangrove forests form an intermediate brackish zone between fresh water and the open sea in many warm areas of the world, including Malaysia. Unlike most tropical forests, there are only a few plant species in mangrove and most of them are trees. The sandy mud under the mangrove trees is soft and deep, and the best way to explore is in a small boat at high tide. The roots of trees which constitute the mangrove forest can tolerate the salinity and lack of aeration caused by constant inundation with sea water. To obtain adequate oxygen, however, mangrove tree roots protrude above the surface of the mud and water, a feature also present in trees of freshwater swamp forests further inland. The mangrove habitat never lacks water or sunlight or essential nutrients. It is not surprising, therefore, to learn that mangrove trees are very productive in terms of rate of growth. Leaves and fruits continually fall into the water and mud below, providing food for crabs and micro-organisms. The waste products of these

Mangrove trees (*behind*) and nipa palms (*front left*) form a link between land and sea in many parts of Malaysia. This vegetation helps to stop coastal erosion and to build new, fertile land by trapping sediment. It is also essential for the reproduction of many fish and prawns.

tiny animals in turn provide food for fish, prawns and molluscs. Apart from the continuous loud chirping of insects, mangrove is relatively still and quiet, but this is deceptive because it is to the mangrove that much of the seafood that we eat owes its existence. The importance, indeed necessity, of mangrove forests in providing feeding grounds and safe nurseries for fishes and prawns in their juvenile stages has become widely realized only in recent years. Indeed, a recent estimate of the value of mangrove-dependent fish and prawn catches in Peninsular Malaysia is well over thirty million US dollars per year. Hardly trivial, but the idea that mangroves are wastelands to be cleared still exists, to the detriment of the fishermen and consumers. The largest tracts of intact mangrove in Malaysia occur on the eastern coast of Sabah, and it is no coincidence that the towns of that region are famous for their seafood. In Sandakan, the export of frozen prawns is partially replacing timber as one of this town's major sources of revenue. Actually, mangrove forests can, if managed with care and restraint, produce wood and seafood permanently. This is a principle known to biologists as 'sustained-yield exploitation'. And this is the reason why the mangrove at Matang Forest Reserve in Perak on Peninsular Malaysia's north-west coast looks like real forest, even after several decades of exploitation for wood, while mangrove cleared only once in certain other areas has turned into an unproductive barren mudflat.

Where sea water receives a frequent flow of fresh water from inland sources, mangrove is usually replaced by dense stands of the nipa palm. These palms often line the banks of rivers on their lower reaches. Nipa fronds once provided the material for roofs in the coastal regions of Malaysia. The large flower buds of nipa can be tapped to yield sugar, but the great difficulty involved in moving around in nipa forest has prevented the possibility of sugar production on a large scale.

Where neither mangrove nor nipa forms the boundary between land and sea, and where there has been no urban development, sandy beaches constitute much of Malaysia's coastline. Wherever there has been a human presence, there are coconut palms, for coconuts are an integral part of Malay cookery, especially on the Peninsula. Along less populated parts of Peninsular Malaysia's east coast, and in the forest behind the beach, look out for the palm-like cycad plant. The cycad is not really a palm, but a bizarre and very primitive plant. It produces seeds on its leaves, without ever going though any flowering stage at all. It is also grown as an ornamental plant in many towns throughout Malaysia. Another plant resembling a palm and which occurs on sand dunes is the screw-pine, a large member of the pandan family, with stiff long leaves fringed with sharp prickles. Leaves of smaller members of this plant family are used in Malaysia to make mats and baskets, and those of one species provide a flavouring for traditional cakes. Trees which resemble pines or some other conifer are often to be seen on low dunes above sandy beaches in Malaysia. Actually, these beautiful trees are not of the pine family; they belong to the genus *Casuarina* and are known as *ru* or *aru*. They are an excellent tree species all round, functioning to reduce erosion, as a windbreak, as a source of high-grade firewood, and they are able to fix nitrogen from the air into nitrate, thus fertilizing the sandy soil.

Freshwater Swamp Forests

Away from the sea, wherever there is inadequate drainage of fresh water or where periodic flooding occurs, there are freshwater swamp forests. There are, in fact, two fairly distinct forms of freshwater swamp forest – one which grows on alluvial soils and the other which has developed on a thick layer of acidic, semi-decayed plant material, known as peat. Swamp forest on alluvium occurs most extensively in eastern Sabah, especially in the floodplain of the Kinabatangan River. The soil under such forest is inherently fertile, but the difficulty in draining it and the risk of massive floods has precluded attempts at converting it to agriculture. Being fertile, the forest is productive of leaves and fruits, although it is neither as tall nor as diverse as the dipterocarp forests further inland. And being productive, the forest supports abundant wildlife, including Proboscis Monkeys, Orang-utans, macaque monkeys, elephants, banteng and waterbirds. Where freshwater swamp forest merges with forest on dry land, there is often a great abundance of strangling fig (*Ficus*) plants. Fig fruits are the favourite food of many forest mammals and birds and, since different plants produce a massive crop of fruit at different times of the year, not during any particular season, there is nearly-constant supply at all times. This is possibly another reason why wildlife is particularly abundant in such regions.

Peat swamp forests have developed most extensively in Sarawak and in some of the coastal regions of the southern half of Peninsular Malaysia. Peat is a highly infertile substrate but these forests, although poor in animal life, are rich in commercially valuable hardwoods and most have been heavily logged. With careful and expensive management, it has proved possible to grow a few agricultural crops on peat after the forest has been cleared. But it is now being recognized that peat swamps provide a valuable service as a sort of gigantic sponge, absorbing flood water which can later be tapped during dry times. Thus, keeping these forests may prevent the need for expensive measures such as dams and water treatment. This means that probably the best way to treat them is by permanent, sustainable logging for timber – an excellent example of the principle that long-term development is best served by being based on conservation.

Dipterocarp Forests

On dry land throughout Malaysia from just above sea level to about 900 metres (3,000 feet) altitude, the natural vegetation is dipterocarp forest. This, then, is the type of rainforest which once cloaked most of Malaysia and which is still the dominant vegetation in the hill ranges where plantations and agriculture have not been established. The strange name – dipterocarp – refers to the fact that most of the largest trees in this forest belong to one plant family, known to botanists as the family *Dipterocarpaceae*. And the meaning of the word (di = two, ptero = wing, and carp = seed) refers to the fact that many of these trees have a fruit consisting of a seed with two wings. Usually, strong winds act to blow the ripe seeds off the tall dipterocarp trees, and the wings help to ensure that they are dispersed away from the mother tree. (In nature there are, of course, many exceptions and the seeds of some dipterocarp species have no wings while others bear five.)

By no means do all trees in a dipterocarp forest belong to this one plant family. Most do not. A recent study at Pasoh Forest Reserve in the state of Negeri Sembilan revealed the presence of 835 tree species in an area of fifty hectares (125 acres), the great majority of which were not dipterocarps. It is worth noting that even an unusually rich forest area of similar size in Europe or North America would contain fewer than one hundred tree species, and none of the climbing plants and palms which are another typical feature of dipterocarp forests.

Having reached a dipterocarp forest, what should we look out for? Let us examine first the tall dipterocarp trees themselves.

Dipterocarp forest, the natural vegetation of Malaysia on dry land from the coastal lowlands to the interior hill ranges. Most of the tall trees, reaching as much as sixty metres (200 feet) in height, belong to a single plant family, the *Dipterocarpaceae*. But the forest as a whole represents one of the most diverse, complex and species-rich systems in the entire natural world.

The trunks of most species are dark in colour, and fissured or scaly. Approach more closely and you will see patches of hard exuded resins on some of the trunks, looking like melted candles or pieces of rock, and known in Malay as *damar*. This damar resin helps the trees to resist infection by fungi, bacteria and insects. Before the widespread development of synthetic substances, damar was collected and exported to Europe, where it was used in the production of paints, varnish and linoleum. And before the advent of battery cells and mineral fuel oils, rural Malaysian people ignited crushed damar mixed with wood for use as lamps. A delightfully fragrant aroma is given off. Nowadays, fresh damar is still used in some rural areas to caulk leaky boats.

Since there are no clear seasons in Malaysia, trees do not show annual growth rings. There is no doubt, however, that large dipterocarp trees are many hundreds of years old. Estimates of the time taken for gaps in the forest (caused, say, by several old trees falling over) to be colonized by plants from the adjacent forest and resemble the original, tall forest again, range from 110 to 375 years. Paradoxically, despite good conditions for plant

growth, the dipterocarps and many other native forest trees grow slowly. One reason for this is that for a sapling or small tree inside the dipterocarp forest, the already tall trees above block out much of the light. Photographers in the forest will find out just how dark it can be in the middle of the day, and scientists have shown that light reaching the ground is only one or two percent of that above the forest canopy. Slowness of growth is also explained by the presence of a vast range of insects, fungi and micro-organisms, ever-present in the forest because there is no dry or cold season to limit their numbers. All tropical forest plants use much of the nutrient that they assimilate not for growth, but for the production of substances which repel the marauding hoards of little organisms waiting to attack. The lignin in wood is one example. Leaves, particularly of dipterocarps, are full of undigestible carbohydrate fibres. Many forest plants contain tannins, which bind chemically to proteins and prevent them being digested and absorbed. Dipterocarps have damar (small amounts occur in the leaves and fruits, as well), which gum up the mouth or the biochemistry of anything which tucks in. Most tropical plants have additional chemicals of varying degrees of bitterness or toxicity which effectively stop animals from eating very much at one go. (Some of these chemicals, incidentally, have medicinal properties, quinine being the best-known example.) All in all, the life of a tropical forest plant is geared as much to preventing itself from being eaten or infected as it is to growth. A third reason why Malaysian forest trees grow slowly is that the natural fertility of soils is generally

Fruits of a wild durian (*Durio* species). This species has not yet been described by scientists.

The base of a strangling fig, showing a system of roots and buttresses which helps to support the massive crown above. Sights such as this are a common feature of the lowland forests of Malaysia.

low. Oddly enough, dipterocarp trees tend to grow most prolifically on hill slopes rather than on the more fertile flat lands.

A feature of many big trees, and small ones too, in Malaysia's forests is the presence of buttresses, like big flanges of wood, at the base of the trunk. It is believed that these act as structural supports. A large proportion of big trees have rotten, partly empty trunks – these can be detected by the presence of holes in the side or base of the trunk, where damar failed to do its job. Such trees infuriate timber-felling companies because they are rejected by overseas timber buyers. They are a delight to the naturalist who realizes their importance as a future seed source, when the non-defective trees have all been felled. The holes are homes and breeding sites for bats, flying squirrels, porcupines and hornbills. Some ecologists have speculated that hollow trees like this are actually better off than their apparently perfect neighbours, because the waste products of animals residing in the holes act as a fertilizer in an otherwise nutrient-poor environment.

In Sarawak and Sabah the Borneo ironwood, known locally as *belian* and also by the splendid scientific name of *Eusideroxylon zwageri*, trails only just behind the South African ironwood in being the densest and hardest wood in the world. At Sepilok, in Sabah, large belian trees can be seen, some of which might easily be a thousand years old. The trunk decays so slowly that it may well last for hundreds of years after the death of the tree. If you walk over the first footbridge along the main trail into Sepilok Forest Reserve, a fallen belian trunk, you may reflect that you are stepping on something which started life before the first European set foot in Asia.

Most people, entering a tropical rainforest for the first time, expect to see plants laden with colourful flowers and succulent fruits. They are often surprised to walk for hours and see only a half dozen or so plants exhibiting either flowers or fruits, and

small, inconspicuous ones at that. This is normal, and the sheer diversity and complexity of plant forms should more than make up for any disappointment. Although weather conditions are similar year-round, many rainforest plants reproduce seasonally. At any one time, it is likely that fewer than two percent of all plants in any dipterocarp forest have either flowers or fruits. Often, but not always and not in all areas, there is a peak of ripe fruits in the forest during the months July to September. In some years, very few fruits are produced at all in the forest as a whole. In other years, quite unpredictably, the forest seems to be filled with all shapes, sizes and textures of flowers and, a few months later, falling fruits. During the past two decades, 1976, 1981 and 1989 have been 'good' years. The majority of dipterocarp forest fruits are tough, bad-tasting and unpalatable to humans, but there are notable exceptions. One is the durian, represented by several species. The durian tree is tall, approaching the size of the dipterocarps when mature, and its fruits are massive, spiky balls containing arrays of large seeds, each coated with a thick nutritious flesh. Only one species – *Durio zibethinus*, with creamy flesh but slightly odoriferous – has been brought into cultivation on a wide scale. The wild, forest species lack the bad odour, but their flesh is less prized by either people or animals. One, *Durio graveolens*, has deep red flesh which tastes like a combination of hazel nuts and avocadoes. There are also various wild relatives of the rambutan (*Nephelium*), mango (*Mangifera*) and breadfruit (*Artocarpus*) in the dipterocarp forest, some with an excellent edible flesh. Few in number but diverse in species are members of the legume family (to which peas and beans belong) – about half are trees and the remainder are woody climbing plants. The thick, green pods of the leguminous climber *Entada*, borne high up in the forest canopy, are a metre in length.

If you make a trip into dipterocarp forest in the hill ranges of the northern half of Peninsular Malaysia, or in Sarawak or Sabah, keep an eye open between 500 and 700 metres (1,600 and 2,300 feet) altitude for a very special plant, *Rafflesia*, which has the largest flower of any plant in the world. *Rafflesia* is a parasite. It has no leaves and it cannot synthesize or absorb from the ground any of the nutrients that it requires. Instead, it is obliged to infect the roots of *Tetrastigma*, a woody climbing plant and member of the grape family. The *Rafflesia* flower takes many months to develop from a small bud to its final size, which varies but sometimes approaches ninety centimetres (3 feet) across when in full bloom. The open bloom lasts for only a few days and so, instead of looking for that, it is better to look in the vicinity of buds or dead flowers. From a distance the buds resemble pale orange cabbages wrapped in charred newspaper, while dead flowers look like pieces of tyre inner tube. The *Tetrastigma* vine is also distinctive, the leaves having toothed edges, like a small saw.

Bears, Bees and Tualang Trees

If we are in forest at a fairly low altitude in Peninsular Malaysia north of Kuala Lumpur or in Sarawak or Sabah, we should look out for the very biggest tree of all. Perhaps surprisingly, it is not a dipterocarp but a legume. Up to about 75 metres (245 feet) tall, with smooth, pale grey bark, and small, pale green leaves, the majesty of the *tualang* tree – scientific name *Koompassia excelsa* and known as *tapang* in Sarawak and *mengaris* in Sabah – has led some people to suggest that this should be adopted as the national tree of Malaysia. Tualang trees are rarely felled in forest reserves and are often left standing when forest land is cleared and burned for agriculture. There are several reasons why tualang are left standing. Their enormous size and buttresses make them difficult to fell, and the wood is hard to saw and to treat with preservatives. Some rural people associate tualang with spirits,

The tualang (*Koompassia excelsa*), largest and most handsome of Malaysia's native trees.

perhaps because of the grotesque woody growths which develop where branches have fallen. Another good reason why rural people like to retain the tree is because it is the favourite of the wild honey-producing bees known scientifically as *Apis dorsata*. Often one or two, sometimes dozens of massive honeycombs can be seen hanging from the high branches of one tree. Obtaining the honey is a very difficult and dangerous job, and with cheap imported honey as well as sugar available even in the most remote shops, the desire and skill to do so is dying out. Even in the past, when honey would have been the only source of concentrated sugars available, few people were willing or able to collect the combs. In Peninsular Malaysia, Orang Asli reach the combs by ascending nearby trees and entering the high branches of the tualang trees along ropes fired with catapults. In Borneo, the method involves the construction of a ladder, by knocking sharpened bamboo pegs into the trunk and progressively lashing them together with rattan cord, so that the load placed on one bamboo step is spread to others. The ascent is slow and laborious, usually done at night, when the bees are unable to see to attack, and it is cooler and less scary for the climber. In rural Sabah, eighty to ninety percent of honey-producing bees' nests are found in tualang trees, even though only a small proportion of tall trees are tualang. Why should this be so? The likely explanation is that Sun Bears, the bees' worst enemy, are not able to climb up into the crown of a mature tualang tree, for the trunk is broad and slippery. Even lianas and other climbing plants are rarely seen in tualangs, when compared to other trees of similar size, perhaps because they too cannot gain a hold on the smooth bark.

The tualang offers one further puzzle of concern to the naturalist. Young trees are found much less commonly than mature trees, and small tualangs are very rare indeed. The flowering and setting of fruit seems to occur rarely, perhaps only once in a decade. The tualang is one of the very few truly deciduous trees in Malaysia. This is a characteristic of seasonal, rather than constant climates. Could the tualang be a relic of an earlier, more seasonal phase in Malaysia's climate, and now

Honeycombs on a tualang tree.

heading for natural extinction? Another curious botanical fact may support this notion. Mangoes do rather poorly in the wetter parts of Malaysia, like much of Sarawak and Sabah, because the flowers need a dry spell to survive and set fruit. Yet recent botanical surveys have uncovered the greatest number of wild mango species, but not the greatest number of trees, in Borneo. This is a paradox: Borneo seemingly the centre for mangoes but with very few wild mango trees in the forest. Could it be that mangoes are another group of trees on their way to extinction where once they thrived?

Latex from the Rainforest

Changing needs and technological advances have caused some once-important Malaysian forest products to become neglected. The giant *jelutong* tree, for example, locally common in some parts of Malaysia, produces a latex suitable for making chewing gum. In decades past, forest trees were tapped for this latex, but a combination of factors led to a decline in use, in favour of latex from an American tree. In some areas, the traces of latex-tapping marks can be plainly seen on forest jelutong trees in the form of decaying, callous-like bark, fifty years or more after the last tappers were active. Jelutong wood is excellent for making pencils. But with the advent of ball-point pens and then computer printers, prospects for bringing this handsome tree into cultivation as a dual-purpose plantation crop for latex and wood are now slight. An attractive woody liana with the scientific name of *Willughbeia* can frequently be found in the forests throughout

Malaysia. It is most often evident when the large, round fruits fall and provide a delicious treat to the forest traveller, for the orange-coloured pulp tastes similar to fresh raspberries. This liana produces a latex very similar to that of the Brazilian rubber tree. Rural Malays used to tap *Willughbeia* rubber for a variety of domestic purposes. When international demand for rubber increased towards the end of the last century, enthusiastic planters tried growing the liana, but its low yield in comparison with Brazilian rubber meant that this native plant was soon out of the commerical race. *Gutta-percha* was the name given to the latex of trees of the genus *Palaquium*, which are found as scattered tall trees throughout the Malaysian dipterocarp forests. The latex is only slightly elastic. It hardens as it cools, and softens when heated. Until the first decades of this century it was in great demand, mainly for insulation of submarine telegraph and telephone wires, and also for making pipes, surgical apparatus, dental fittings and golf balls. The lesson to be drawn from these examples is that the dipterocarp forest has much more to offer than either timber or interesting wildlife. It is, most importantly of all, a living store of natural products, the value of which lies largely unrecognized. Who knows which plants will provide the most benefits one hundred years hence?

A stream on the fringes of dipterocarp and heath forest. The water is clear but tinged reddish-brown, a result of percolating through a leaf litter rich in tannins and other natural chemical compounds, and a typical feature of tropical forests on infertile, sandy soils.

Heath Forests

In scattered parts of Malaysia, heath forests may be found, consisting of low, stunted trees with generally small, thick leaves. Very few animals live here, even insects, although rhinos and Barking Deer sometimes pass through to browse upon a few new leaf shoots. More than one biologist has remarked that a deafening silence is the feature which distinguishes this forest type from other tropical rainforests. Heath forests may be found from near sea level up into the mountain ranges, but they are never extensive and are normally situated on flat terrain or on ridge tops. They occur wherever the soil is sandy and extremely deficient in nutrients. Often they are demarcated by a clear, sharp boundary from adjacent dipterocarp forests, but in some cases they merge progressively. Due to the paucity of nutrients, no plants can grow very large and instead species with special characteristics have evolved to cope with these conditions. Apart from thick leaves, which help to minimize attacks by foliage-eating insects and mammals, a feature of some heath forest plants is their mutually beneficial association with ants. *Hydnophytum*, for example, grows on the trunks of trees, and has hard, bulbous tissue which contains chambers which serve as a home for ants. In return for this ready-made nest, the ants reward the plant with protection from caterpillars, plus nutrients in the form of their waste products. Other heath forest plants with different forms have a similar symbiotic relationship with ants. Pitcher plants, which we shall meet again on the similarly nutrient-poor soils of high mountain slopes, also abound in heath forest. The pitchers are not flowers or fruits, but a specially-evolved structure which serves to catch and digest insects and other small animal life, a much-needed source of elements like nitrogen and phosphorus.

Montane Forests

If we walk in dipterocarp forest from the lowlands up into the hill ranges and mountains, subtle and very slight changes in the forest's structure and composition become apparent with increasing altitude above sea level. For example, the density of large trees increases, but very large ones, like the tualang, cease to occur; orchids become more common; trees with very large leaves become rare. Then, at a certain point, a fairly sharp change will be noticed and it is here that we have entered montane forest. The exact altitude at which the change occurs varies from place to place. On small, isolated mountains, it may be as low as 600 metres (2,000 feet), while on the massive mountain ranges of western Sabah it may be above 1,200 metres (4,000 feet). Very large trees and large lianas cease to exist. There are few or no dipterocarps and legumes. Instead, trees of the oak (*Fagaceae*), myrtle (*Myrtaceae*) and laurel (*Lauraceae*) families suddenly dominate the forest. There are also magnolias, rhododendrons, raspberries and other plants which seem out of place in the tropical zone. Boulders carpeted with moss appear on the ground, and tree trunks as well as branches become covered with mosses, orchids and other epiphytic plants. Pitcher plants of several species are common. There are fewer bird and insect noises, and they are different from those which have become familiar lower down the mountain. The forest is often suddenly enveloped in mists – bird and insect sounds cease, visibility drops to less than ten metres, and the forest becomes cold and eery. With increasing altitude, all these features become increasingly more pronounced. A different array of plants takes over. Trees are replaced by gnarled bushes. Lianas disappear, and even the mosses and orchids drop out as the amount of space on which they can survive decreases.

LEFT An epiphytic plant (*Hydnophytum formicarium*) on a small tree in a Sarawak heath forest. The expanded bases of the stems of this plant contain chambers in which ants dwell. This arrangement provides nutrient and protection from caterpillars for the plant, in return for a ready-made house for the ants.

BELOW Wild raspberries growing on Mount Kinabalu at 2,100 metres (7,000 feet).

Animal Life

The forests of Malaysia possess a rich variety of animal life. But while there are a great many species, the actual number of individual animals is relatively small and they are often difficult to locate. One obvious reason why larger animals are seldom seen in the rainforest is, of course, that with so many trees everywhere, it is just not possible to see any creature on the ground if it is more than about twenty metres away. Animals in the tall tree-tops can be seen up to about seventy metres away on average, but that is still not very far. It is because they are tree-dwellers that monkeys are amongst the most easily and commonly encountered of rainforest wildlife, coupled with the fact that they are fairly noisy and active during the day. It must be remembered, too, that all animals, ranging from ants to elephants, remain hidden whenever they can, often sitting or standing about doing nothing in particular for hours on end, rather than wasting energy and abandoning safety by running conspicuously around the forest.

There are various scientific reasons why large animals are rare in wild Malaysia even under entirely natural conditions. One important factor is the array of parasites and predators and diseases, nurtured in a constantly warm, moist environment, waiting to keep in check any form of life – large or small – that becomes more than usually abundant. A further consideration, and the main reason why there are no wildlife spectacles like those on the plains of Africa, is that rainforest plants are protected from leaf-eating mammals by indigestible and toxic chemicals. Animals like deer, elephants and wild cattle require large quantities of succulent herbage, something which hardly exists under tall forest conditions, where the ground is instead carpeted with slow-growing woody saplings and spiny palms. Members of the antelope family are totally absent from the Malaysian region. A lack of large grass-eating animals in turn means that large predatory cats are rare or absent. In Peninsular Malaysia, the Tiger, Leopard (usually the black variety of the spotted species) and Clouded Leopard occur but never in abundance, and the Tiger is now very rare indeed. In Sarawak and Sabah, only the Clouded Leopard is present. Stories of other large cats in these states invariably turn out to be false or, at best, highly dubious. The fact that many large animals, including elephants, rhinos, tapirs, wild cattle, deer, pigs and leaf monkeys visit natural mineral springs and salt licks has long been known to hunters and naturalists. It is clear that they do not merely like to eat the mud or drink the water, but are obliged to for their health and survival. It seems probable that some soils in Malaysia contain inadequate minerals for large animals, and this is another reason why their numbers are limited.

The lack of very large mammals in Malaysia may also be explained by the hot, humid climate and by evolution. Put simply, large animals in a hot climate tend to overheat. In dry heat, they can help to cool themselves by losing water through the skin in the form of sweat, which evaporates, taking heat energy with it. In Malaysia, the air inside the forest is saturated with water, and instead of evaporating, sweat accumulates on the skin. Large wild mammals cope in various ways. They move about and feed mainly at night and during the early hours of the morning, when it is coolest. Rhinos wallow in muddy pools, where possible on breezy hillsides. Tapirs seek rivers. Deer lie up in thick shade. Elephants disappear very mysteriously but presumably do the same as deer. On an evolutionary timescale, we can see indi-cations of the advantage of small size in a hot, humid climate. Malaysian elephants, rhinos, tigers and bears are the smallest members of their kinds in the world. Fossils from Niah in Sarawak and from elsewhere show that rhinos, apes and some other animals have become smaller in size in a matter of only tens of thousands of years, corresponding with the development of a hotter, wetter climate. Amongst people, natives of Malaysia are smaller than visitors from temperate countries, and amongst Malaysians the forest-dwelling Orang Asli and Penan tend to be the smallest. There are two other reasons why large mammals are rare in Malaysia. But both have to do with the recent exploitation of the natural environment and I shall return to them under *Wildlife Conservation*.

In the following section, *A Walk Through the Rainforest*, some insights will be offered into forest life by describing what we are most likely to encounter along a forest trail. Some large animals can be seen fairly easily. More often, strolls may have to extend into days of extensive tramping, in order to be rewarded with the sight of some of the more spectacular animal species for which Malaysia is renowned. Let me now describe to you some of the larger wildlife species and also provide some tips as to how you might go about finding them in the wild.

Elephants

The Asian Elephant (Malay: *gajah*), as the largest land animal, deserves to come first. Adult Malaysian bull elephants are roughly 2.5 metres (8 feet) tall, while cows reach about 2.2 metres (7 feet). These sizes are smaller than those of the same species of elephant in India, and much smaller than those of African elephants. The elephant is probably the most endangered animal species in Malaysia basically because just one herd requires a very large area in which to survive. They are very selective in what they eat – grasses, the growing tips of palms, stems of wild bananas and gingers make up the bulk of their diet. Very few woody plants are consumed. It is a common belief that elephants cannot live in the hills, but this is not strictly true. They are surprisingly adept at negotiating steep slopes, but rarely spend much time in the hill forests, probably because of the lack of suitable food and mineral salts there. As the original forest habitat disappears to make way for plantations and towns, herds have become separated and isolated into different reserves and parks. Fresh genes cannot enter into separated herds, which suffer increasingly from a greater risk of extinction. For the present, however, elephants survive in scattered parts of Peninsular Malaysia and in eastern Sabah. There are no records of wild elephants ever having occurred in Sarawak or in western Sabah. Some people believe that elephants were introduced into Borneo by a timber logging company. This is indeed so, but the species was already present when Europeans settled in Sabah (then British North Borneo) in the last century, long before any logging started. Another story is that a small number of elephants were released on the east coast of Sabah several hundreds of years ago, probably by the Sultan of Sulu. While this may indeed have happened, it is highly unlikely that such a small group could have built up to number the 2,000 or so elephants believed to have been present in Sabah by the end of the last century. In time gone by, even up to the early years of this century, families of high status in Kedah, Perak, Melaka, Terengganu and Kelantan possessed tame elephants, caught locally from the wild, which were used both as a status symbol and as a means of transportation over long distances. So it is possible that tame elephants were once released and have returned to their wild state in Sabah and elsewhere. Whatever the true explanation for the absence of elephants in western Borneo, it is now the case that eastern Sabah has the greatest remaining concentration of this species in Malaysia. It is difficult to travel in eastern Sabah without one's own four-wheel-drive vehicle, but enthusiasts who wish to see wild elephants could head for any of the new oil palm plantations bordering on to forest in this region, preferably during rainy weather, when elephants tend to travel far outside the forest cover. Alternatively, they could hire a small boat on the lower Kinabatangan River, preferably during a dry spell, when elephants do not venture far from forest cover near a permanant water source. In this case, one should ask to be taken up any of the small side streams, especially those which lead to oxbow lakes, where there is a chance of seeing elephants in the early morning or late afternoon. These areas are also rich in Proboscis Monkeys, Orang-utans, otters, waterbirds and other wildlife. Elephants are more scattered in Peninsular Malaysia; they occur in and around Taman Negara, and in the proposed Endau-Rompin Park, but seeing them is largely a matter of luck.

The structure of groups is the same as that found amongst elephants elsewhere. Herds of approximately ten to twelve are commonly seen and they consist of several adult females with their young. These herds often split up into smaller units which forage separately. Mature bulls live a basically solitary existence. They range up into steep hilly areas more frequently than the cow-calf herds, but often follow the herds around in order to mate. Occasionally several herds may merge and between thirty to fifty elephants roam together through forests and plantations.

Wild Cattle

The second largest land mammals are two kinds of wild cattle. In Peninsular Malaysia there is the *seladang* (known as *gaur* in India), while on the island of Borneo occurs the smaller *banteng* (widely known as *tembadau* in Sabah). The bulls of both species are much larger and more powerfully built than cows. Seladang are black with creamy-white markings like stockings on the legs. Banteng are less massive and are distinguishable by their white rump; the body of the adult bull is black, but that of females and young males is a pale reddish-brown. Both species are, like other cattle, basically grazers. They can live in the forest but find little food there, and it is possible that they have survived into this century only because some areas of forest have been cleared by people for planting hill rice. With the widespread use of guns in recent decades, however, numbers have been decimated in many areas, and much potential grassland habitat goes unused. Like elephants in Malaysia, herds of seladang and banteng are scattered and isolated by plantations and roads. Seladang are most likely to be found in Taman Negara and other forest areas in Pahang and Kelantan, but they are shy and rarely seen. Much the same applies to the banteng; like elephants, they are most likely to be encountered on rough roads in the lowlands of eastern Sabah. Both species often visit natural mineral sources. There has been considerable interest in recent years in the possibility of domesticating seladang and banteng for meat production. This idea makes sense, because both species are better adapted to local conditions than are any of the usual breeds of domesticated cattle, which originate from Europe and India.

Rhinoceroses

Malaysia is home to the world's smallest rhinoceros, usually called the Sumatran or Asian Two-horned Rhinoceros (Malay: *badak*). Some people prefer the appropriate name of hairy rhinoceros, although bristly is perhaps a more accurate description. This rhino is now very rare and until a few years ago was feared extinct in Borneo and in many parts of Peninsular Malaysia. Surveys since the late 1970s have revealed that it is present, in very low numbers, in many parts of the country. Fearing that the wild populations may be in danger of extinction,

A bull seladang (*Bos gaurus*). Now mainly confined to the remoter forest regions, these majestic animals may sometimes be seen when they visit natural mineral licks or at the forest fringe, where they come to feed on grasses at night.

The Sumatran Rhinoceros (*Dicerorhinus sumatrensis*), a rare, shy inhabitant of the extensive forest regions of Malaysia.

the governments in Peninsular Malaysia and Sabah have started programmes to capture rhinos isolated in unprotected forest areas and to bring them together in captivity in the hope of establishing a managed breeding population. It would indeed be particularly tragic if this animal were permitted to become extinct, because of all large living mammals, it retains the distinction of having been around almost unchanged for about thirty million years, longer than any other. Another remarkable feature of this rhino is its diet: it is the only mammal (apart from the tapir, described below) to live almost exclusively on the mature leaves and twigs of woody forest bushes and saplings. How rhinos and tapirs can cope with such a tough diet, but not other mammals, is a mystery. The surprising thing about the hairy rhino is that it *is* so rare, for the forest seems to be full of rhino food. Presumably, relentless hunting over hundreds of years is largely to blame, although the scarcity of minerals in the rainforest environment may be a contributory factor. The reason why persecution of this species is so severe is that Chinese communities throughout Asia have for long believed that rhino horn can cure fevers and other illnesses. A very small amount of horn is believed to be effective, and in recent years the price of hairy rhino horn has exceeded that of the same weight of gold. The horn of the hairy rhino is reckoned to be the most effective of all the species. Western medicine does not recognize rhino horn as an effective remedy as is claimed by the Chinese, but controlled experiments to test the claims have never in fact been carried out. The situation now is that hairy rhino horn is smuggled from the countries of origin to Chinese communities elsewhere and sold on the black market. Conservationists claim that rhino horn has no medicinal properties and use this as the argument to try to stop trade in the horns. But beliefs of the Chinese are firmly ingrained. Perhaps a better approach would be to test the properties of rhino horn, admit that it may be effective (if only psychologically) and on that basis argue the importance of saving the species from extinction. Both sides would then be fighting for the same cause.

In their natural habitat, the presence of rhinos is most often detected by their footprints. There are three large toe-nails on each foot, one in front and one on each side, with a maximum width between the side nails of between eighteen and twenty-four centimetres (7 and 9½ inches). The hairy rhinoceros is normally a solitary animal, although pairs and groups of three are often reported in Sabah. Another sign of their presence is where mud wallows have been dug into the soil. Rhinos often spend the hot hours of the day enjoying a cooling mud bath here. Experienced hunters know this, and wait concealed for days or weeks until their prey arrives. Wild pigs also make such mud wallows, but these can be distinguished by the presence of numerous cloven hoof prints, rather than a few large three-toed prints. In Peninsular Malaysia, there is a chance of seeing wild rhinos at Sungai Dusun, Ulu Selama and Endau-Rompin, but these places are normally not open to the public. In Sabah, Tabin Wildlife Reserve and the Danum Valley conservation area are the main rhino areas, but here also access is difficult. Let us hope that the new captive breeding programmes will permit people to see the wonderful hairy rhinoceros in perpetuity.

Tapirs

Related to the rhinoceros and found in Peninsular Malaysia, but not in Sarawak or Sabah, is the Malay Tapir. This strange, smooth, black-and-white mammal is smaller than the hairy rhino but larger than a wild pig. It has a long, flexible snout, almost worthy of being called a trunk. Its footprints are similar to those of the rhino, but smaller. It is not regarded as having medicinal properties, although its diet is similar to that of the hairy rhino, and it can probably survive in a fairly small area of forest. Thus, in Peninsular Malaysia, it is more common and widespread than the rhino. Tapirs are still sometimes seen as close to Kuala Lumpur as the Ampang Forest Reserve, although the largest surviving population probably occurs in and around Taman Negara.

Serows

If you go into the forested hills of Peninsular Malaysia and glimpse what appears to be a blackish-coloured pony miles from the nearest human habitation, you have probably seen something called the Serow. The Serow (Malay: *kambing gurun*) is actually a relative of the goat. Both sexes bear a pair of almost-straight horns twenty to thirty centimetres long (8–12 inches), but these are sometimes difficult to see from a distance. This animal is often associated with limestone hills, but it also occurs far from limestone, in such places as Fraser's Hill.

Pigs

There are two species of wild pig, both known locally as *babi hutan*. One of them, the Bearded Pig, was once widespread throughout the region, but in Peninsular Malaysia it has largely been replaced by the adaptable Common Wild Pig, the same species that extends in distribution to Europe, northern Africa, mainland Asia and Japan. The Bearded Pig takes its name from a fringe of long, bristly hairs on the jaws. It is the only wild pig species in Borneo, where it has adapted its way of life to extensive dipterocarp forests and mountain forests dominated by trees of the oak family, through which it roams in search of fallen fruits. The Bearded Pig is a major source of meat for hill-dwelling peoples, especially in Sarawak. Both species of wild pig will invade plantations and gardens after dark in search of food. In the rural villages of Borneo, one may encounter grotesquely fat, short-legged black pigs, but these are of a domestic variety.

A mother Wild Pig (*Sus scrofa*) with one of her piglets. This animal is equally at home in deep forest, or on the fringes of cultivation, which it may enter to feed on root crops or fallen fruit.

Deer

Malaysia has five kinds of deer. The largest is the Sambar Deer (Malay: *rusa*), each antler of a fully mature stag bearing three points. This deer prefers forest edges and river banks, where it feeds on herbs and grasses. There are two species of Barking Deer (Malay: *kijang*), one of which occurs only in Borneo. Males have short antlers, females have none. Both species prefer to stay under the cover of forest, where they feed on fallen fruit and leaf shoots. The name 'barking' refers to the loud, hoarse barks made by these otherwise reclusive deer, a noise which can cause considerable alarm to the unwary. Barking Deer can sometimes be seen while driving along roads in the hill forests, or during a walk in hilly terrain. Their seeming dislike of flat land is strange. Rural hunters are able to call Barking Deer towards them by making a high-pitched sound with the aid of a leaf. This sound is said to resemble the call of a baby Barking Deer. There are also two species of Mouse-deer, in which neither sex bears antlers. The Greater Mouse-deer is known locally as *napu*. The Lesser Mouse-deer, less than thirty centimetres (1 foot) in height and known as *pelandok* or *kancil*, features in several traditional Malay stories. Unscrupulous sellers of traditional medicines and charms sometimes display 'mouse-deer' horns, which are, of course, fakes and are often the long canine teeth of the males. Mouse-deer can often be seen at night in the lowland forests. Although also periodically active during the day, their small size makes them difficult to spot. Rural hunters are sometimes able to attract Mouse-deer by making a drumming sound with two small sticks on a large, dry leaf. This is said to resemble a contact noise made by these little animals when they stamp their hind feet on the ground.

The Common Barking Deer (*Muntiacus muntjak*), a rather small, graceful forest deer which has a surprisingly loud barking call. It is a common inhabitant of all the hill ranges, where it feeds on herbaceous plants and fallen fruit.

Monkeys and Apes

For those with an interest in mammals, but with not enough time to spend searching for those described so far, efforts to see monkeys and apes are likely to be repaid. These animals – which together with the Slow Loris, Tarsier and *Homo sapiens* belong to the group of mammals called primates – are not only active in the daytime but also fairly noisy and conspicuous.

South-east Asia is the home of the leaf monkeys, also known as langurs, and Malaysia has seven species. They feed mainly on the seeds and leaves of forest trees and lianas, especially those of the legume family. A trip to Kuala Selangor provides a good chance to see the wild but fairly approachable Silvered Leaf Monkey. Adults are the colour of steel wool, babies a brilliant orange. Away from the coast in Peninsular Malaysia, not only in forests but also in abandoned rubber plantations around Kuala Lumpur and Petaling Jaya, we can see the Dusky Leaf Monkey (named after its sooty coat, and also known as the Spectacled Langur, because of the white patches around its eyes) or the Banded Leaf Monkey (a greyish monkey which cannot by any stretch of the imagination be said to be banded). Both species are abundant in many areas, but they are particularly conspicuous at the Kerau Wildlife Reserve in Pahang. In the past, and to a very much lesser extent now, these monkeys were the main protein source of many Orang Asli, who hunted them with blow-pipes and poisoned darts. There occurs in the forests of Sarawak a species closely related to but different from the Banded Leaf Monkey. In Samunsam Wildlife Sanctuary, it is almost entirely black, while in northern Sarawak it is a mixture of reddish-brown, black and white. The Red Leaf Monkey can be found throughout most of Sarawak and Sabah. Similar in colour to the Orang-utan, it is often confused with that ape by those who are unaware that monkeys have tails while apes do not! This species may be seen at Sepilok and in the Danum Valley conservation area in Sabah. In northern Sarawak and Sabah is the Grey Leaf Monkey, a handsome grey-and-white animal which is often detected not by sight but by the bizarre gurgling call of the adult male, which mystifies even some very experienced Borneo travellers. It is common in the hill ranges of north-western Borneo and in Tabin

An infant Silvered Leaf Monkey (*Presbytis cristata*), a scarce monkey of Malaysia's coastal fringes. This youngster, born bright orange in colour, is already beginning to change to its adult steely grey.

Wildlife Reserve in Sabah. In south-eastern Sabah, including Danum Valley and Tawau Hills Park, there are creamy-white leaf monkeys. These animals still await classification by zoologists. Lastly, in the Lanjak-Entimau Wildlife Sanctuary and other parts of interior Sarawak is the White-fronted Leaf Monkey, the least-known of all the species. Many centuries ago, a highly sought-after cure for various illnesses was bezoar stone. This is a greenish-coloured concretion of a tannin compound which accumulates over a period of years in the stomachs of some herbivorous animals. The major source in South-east Asia was leaf monkeys. In fact, the substance probably has no medicinal properties and the value formerly placed upon bezoar stones is now all but forgotten.

Closely-related to the leaf monkeys is the Proboscis Monkey, a unique creature found only on the island of Borneo. Mature males are enormous, commonly weighing over twenty kilograms (45 pounds), and having large, pendulous noses. Mature females are only half the size, with a small snub-nose. For some reason this monkey species chooses to live near water. It occurs in the coastal forests of Sarawak and Sabah, including mangroves, and may be seen tens or even hundreds of kilometres inland along the banks of rivers, as well as in the swamp forests adjacent to the larger rivers. Despite its rather fearsome appearance, it is a very gentle monkey, feeding only on plants. In the late afternoon, good views may be had of large groups of them along the lower Kinabatangan River in eastern Sabah and also at Samunsam Wildlife Sanctuary in Sarawak.

Much less attractive, and regarded as serious pests in plantations and rural gardens are the Pig-tailed and the Long-tailed Macaque monkeys. The former is a creature of hill forests. In Kelantan and Terengganu, young male Pig-tailed Macaques are caught and trained to pick coconuts: you may see them with their owners when travelling in these states. The latter is an adaptable

ABOVE The Dusky Leaf Monkey (*Presbytis obscura*).

Profile of a young male Proboscis Monkey (*Nasalis larvatus*), a unique primate which occurs only in some coastal regions on the island of Borneo. Later in life, with a enormous, cucumber-like nose, this individual will be master of a harem of up to ten or so mates.

Portrait of a Long-tailed Macaque (*Macaca fascicularis*), with characteristically prominent cheek whiskers. These monkeys are common in coastal forests, including mangrove and beach, also around villages and plantations where they sometimes raid vegetable gardens.

monkey, most commonly seen on the fringes of towns, villages, cultivated areas and forests, and along the banks of large rivers. It is also known as the Crab-eating Macaque, because groups resident in coastal areas subsist largely on crabs and other marine organisms.

Gibbons – small, agile apes – are widespread throughout Malaysia and the high-pitched songs of the females are one of the characteristic sounds of the Malaysian dipterocarp forests. Gibbons are amongst the very few mammals which live in monogamous family groups – male, female and their offspring – with the parent animals generally pairing for life. The closely-related but larger Siamang, totally black in colour, occurs only in the hill ranges of central Peninsular Malaysia, where it is one of the most memorable sights and sounds of the forest.

Perhaps most appealing of all the primates is the Orang-utan, confined in Malaysia to parts of Sabah and Sarawak. The name 'orang-utan', incidentally, is an invention, meaning 'forest person'; in Sarawak they are called *maias* and, in Sabah, *kogiu*. Orang-utans are strange animals when compared to the other primates. Arguably the most intelligent land mammals after human beings, they are the least social of all apes and monkeys.

Individuals living in the same part of the forest know each other and meet from time to time while sharing breakfast in a fruit tree, but for most of the time they ignore each other and go their own way. Each Orang-utan above the age of roughly four years makes a new nest for itself every day, in which it sleeps throughout the night. The method used is simple: a forked branch is found in a tree, anywhere between five and fifty metres (16 and 160 feet) above the ground, and adjacent small branches and twigs are bent and snapped towards the fork to form a roughly circular platform. With the possible exception of the Sun Bear, no other animal makes nests like this, and so they are a useful indication of the approximate whereabouts of Orang-utans. The animals themselves are normally very quiet and difficult to spot in the wild. It is quite possible to travel through the forests of eastern Sabah and see dozens of their nests, some freshly made, but only occasionally does an Orang-utan show itself. It was once thought that the favourite abode of Orang-utans are tall, extensive dipterocarp forests. Recent studies have shown that they are generally rare or absent in such forests, and instead occur most abundantly in swamp forest, and also in forests near the coast and along rivers.

LEFT The Orang-utan (*Pongo pygmaeus*), the largest and most intelligent Asian primate (apart from human beings), occurs in the Malaysian states of Sabah and Sarawak. Most easily viewed at Sepilok Forest Reserve, the bulk of the wild population lives in thick swamp forests.

BELOW Perfectly adapted to life in the trees, gibbons (here the Bornean Gibbon, *Hylobates muelleri*, from Sarawak) are a characteristic feature of the dipterocarp forests throughout Malaysia. Swinging like pendula beneath long arms, gibbons scour the forest for juicy fruits, tender young leaves and insects.

RIGHT The Slow Loris (*Nycticebus coucang*) is an appealing, soft-furred primate which sleeps in a secluded spot above the ground during the day, becoming active after dark. It may be found in rural gardens and plantations, as well as in the rainforest.

BELOW The Tiger (*Felis tigris*) occurs in Peninsular Malaysia but not in Borneo. Now an endangered species in the country, this magnificent cat rarely comes into direct conflict with people, but occasionally takes domestic cattle in rural areas.

The Slow Loris and the Tarsier

Discussion of the primate family would be incomplete without a mention of two little animals which become active only at night. One of them is the Slow Loris (Malay: *kongkang*), a rather tubby, soft-furred creature with a very short tail. When frightened, the Loris is not as slow as you might expect. It is not rare, and indeed is often encountered in wooded areas around towns and villages throughout Malaysia. In Sabah, it has adapted well to life in cocoa plantations, where it feasts on the cocoa fruits. To make up for this behaviour, it also consumes many insect pests, and plantation managers seem satisfied with this trade-off. In Sarawak and Sabah, but not in Peninsular Malaysia, is one of the strangest of all mammals, the Tarsier. Only about thirteen centimetres (5 inches) long, excluding a pencil-like, twenty-centimetre (8-inch) tail, the Tarsier has soft fur, enormous round eyes, and frog-like hands and feet, which bear nails on some fingers and toes, and claws on others. It can rotate its head through 180 degrees to face very nearly back to front. If you shine a flashlamp at the Tarsier's face, the eyes do not reflect light, unlike the great majority of animal species active at night. The Dusun people of Sabah call this animal *tindukutrukut*, a name which refers to this curious mixture of features.

Cats

There are several species of wild cat in Malaysia including the Tiger, Leopard and Clouded Leopard. The attractively-marked Leopard Cat, pale yellow with black spots, is about the size of a domestic cat. It has adapted well to human activities and may be seen in plantations and on the fringes of villages, feeding on insects, rats and occasionally chickens. In contrast, the Bay Cat is

known only from the forests of Borneo, where several specimens were collected for museums during the last century. Since then there have been no definite sightings of this cat, although there are reports of its existence near the Danum Valley conservation area in Sabah. Anyone who can confirm the continued existence of the Bay Cat will have made one of the major zoological rediscoveries of modern times.

The Sun Bear, Binturong and Other Carnivores

The Sun Bear, or honey bear, is the only kind of bear in Malaysia. Whether it is as partial to honey as some people believe, there is no doubt that the contents of bees' and termites' nests make up the bulk of its diet. This bear can climb quite well, but is too heavy and ungainly to jump from branch to branch, so it has to make do with trees that are easy of access. They are active during the day and at night. Rather similar to the bear, but belonging to the civet family is a powerfully-built animal with a long tail and thick bushy black fur called the *binturong*, or Bear Cat. Other small carnivorous mammals which occur in Malaysia include the Yellow-throated Marten, the Malay Weasel, various species of otter, civet and mongoose, the Hog-badger (in Peninsular Malaysia), Ferret-badger (confined to Mount Kinabalu) and the Malay Badger, or *tudu*, also known as *teledu* (in Sarawak and Sabah).

ABOVE The Sun Bear (*Helarctos malayanus*) is a rarely-seen denizen of the Malaysian rainforest, occurring from the coastal swamps to the high mountain ranges. Its powerful claws allow it to rip open soil and wood, while its long tongue helps to mop up the termites and ants on which it feeds.

BELOW A member of the civet family, the thick-furred *binturong* or Bear Cat (*Arctictis binturong*) has a strong, prehensile tail which enables it to live up in the tree-tops. It is fierce and partially carnivorous, but has a particular liking for the fruits of strangling figs.

The Common Palm Civet (*Paradoxurus hermaphroditus*), one of the commonest members of the civet family. It has no particular liking for palms, however, and may be seen searching for food on the ground and in low trees in disturbed forests and rural gardens.

Squirrels, Treeshrews and the Flying Lemur

Malaysia has twenty-six species of squirrel which are active during the day, ranging in size from the Black Giant Squirrel (*Ratufa bicolor*) of Peninsular Malaysia to the tiny twenty-gram (³⁄₄-ounce) pigmy squirrels of Borneo. There is a fairly distinct array of those species which live in the lowland forests, and another array which are confined to hills and mountains. Curiously, the distribution of the mountain species starts in the upper dipterocarp forests, below montane forest. With so many species, squirrel-watching in Malaysia can become nearly as much of a hobby for the amateur naturalist as bird-watching. The treeshrews – superficially resembling small squirrels – are a group of mammals confined to Asia, which have a long, pointed snout, feeding on insects and small fruits. They are not closely related to any other living mammal group, but in physical appearance resemble the very earliest mammals evolved from reptiles during the age of the dinosaurs. Of the world's sixteen treeshrew species, ten occur on the island of Borneo. The Mountain Treeshrew (*Tupaia montana*) is the commonest mammal in Kinabalu Park, where it may frequently be seen during a morning walk on the trails.

At night, flying squirrels (which actually glide with the aid of membranes between the front and back legs), rats and porcupines take over. If, as dusk is drawing near and you are in a lightly-wooded area or at the forest edge anywhere in the lowlands of Malaysia, you glimpse out of the corner of your eye what appears to be a large tray shooting silently through the air, look again and you will probably have seen a Red Giant Flying Squirrel about to embark on its nightly search for food in the form of seeds and new leaves. This splendid animal is nearly one metre (3 feet) long, half of which is its bushy, black-tipped tail. It can make continuous glides of over one hundred metres (330 feet) between two trees. An animal called the Colugo, or Flying Lemur, is rather similar in appearance to the Giant Flying Squirrel. It differs from the flying squirrels in many details, including having the gliding membrane extending from the hind legs to the tip of the tail.

The Mountain Red-bellied Squirrel (*Callosciurus flavimanus*). This species occurs in the mountain ranges of Peninsular Malaysia, feeding primarily on fruits and insects.

The Dayak Fruit Bat (*Dyacopterus spadiceus*) is the largest Malaysian bat after the massive flying foxes.

Bats

Nearly one-half of the mammal species in Malaysia are bats, all of which become active around dusk and normally fly only during the night. They are the only mammals capable of true flight. There are actually two distinct groups of bats, which scientists believe probably evolved from two completely different ancestral lines. The Megachiroptera group range in size from small to large and all feed on fruits, flowers or nectar. The Microchiroptera group, to which most species belong, feed primarily on flying insects. Microchiropteran bats navigate and find their prey not by sight but by echolocation – that is, by sending out high-pitched sounds through their mouth or nose, and gaining an image of their environment by the pattern of echoes which return. Although small, these bats are the most numerous of the Malaysian mammals. No one has made any calculations, but it would not be surprising if there are hundreds of millions of bats in the country. What a lot of insects they must eat!

Blood-sucking bats do not occur in Asia, but the enormous Flying Fox with a one-and-a-half metre (5-foot) wingspan, is actually a bat although it is sometimes incorrectly referred to as a vampire. It eats fruit and causes great damage to orchards throughout Malaysia. Flocks of hundreds or even thousands may be seen flying after dark into orchards when the fruit is ripening. At other times of the year, these giant bats spend much of their time in coastal regions, and great roosts can sometimes be found in the mangrove forest.

Birds

Malaysia is particularly rich in bird life and an experienced ornithologist can identify by sight and sound perhaps as many as sixty species during a day's walk in the lowland dipterocarp forest. Birds are also more readily seen and heard than mammals. A good way to see many colourful species is to find a large strangling fig plant in fruit, and wait in a concealed position with a view of the crown. Strangling fig plants can be recognized by their net-like tangle of woody roots, usually smooth and grey, sometimes reddish tinged, which enclose and eventually kill the entire tree on which they started their life. Old stranglers may be over fifty metres (165 feet) tall. The numerous yellow or reddish

The Wreathed Hornbill (*Rhyticeros undulatus*) occurs in the extensive forested hill ranges of Malaysia, where it is often first detected by its sharp, barking call and heavy, rhythmic wingbeats. The bright yellow throat pouch is characteristic of the species.

The Striped Bronzeback (*Dendrelaphis caudolineatus*), one of Malaysia's numerous handsome and harmless snakes. Forest snakes such as this are shy, uncommon and rarely seen. They feed primarily on small vertebrate animals.

fruits, borne in the massive leafy crown, are a favourite food of hornbills, barbets, pigeons and many other birds, as well as mammals. Hornbills are generally conspicuous because of their loud calls, large size and noisy wing-beats when in flight.

Another good place to observe birds is in a secluded valley in the depths of the dipterocarp forest. Listen for bird calls. Imitate them if you can, and the bird may advance towards you. Failing that, move slowly and soundlessly towards the calls. A long, rather low-pitched monotonous whistle will usually reveal itself as the Garnet Pitta, a ground-dwelling bird of damp, shady spots, which feeds on insects, grubs and snails. Various pheasants and partridges also forage on the forest floor, where they find fallen fruits and insects as food. The source of a sweet-sounding 'chee-cheecheechee' from the canopy of low trees will at first be invisible, but closer inspection may reveal the beautiful Green Broadbill. Solitary flashes of red and black in low forest trees will probably belong to a species of trogon, while groups of red and black birds flying among the tree tops will be minivets. Eagles, hawks and other daytime birds of prey are infrequently seen in the forest, because visibility is poor, but their piercing calls may be heard, sending tree squirrels scurrying in alarm for cover. At night, a variety of owl calls may be heard, and sometimes even during the day an owl is accidentally flushed from its resting place by our presence.

Snakes, Lizards and Crocodiles

Well over one hundred species of snake occur in Malaysia, of which less than twenty are at all poisonous, and only two, the King Cobra and Black Cobra, are likely to cause the death of human beings through their bite. The cobras are at least as common in plantations and villages as in the forest. Indeed, snakes are much more often encountered on roads and in gardens than in the forest, where one can walk for days without seeing one. Small lizards occur nearly everywhere, and are especially noticeable on forest edges. The Water Monitor Lizard (Malay: *biawak*), which may grow to two metres (6½ feet) in length, lives near rivers and in swampy areas. At a glance, it may be mistaken for a crocodile, but differs in having a shorter snout, a rounded rather than flattened body, and a long, slender tongue which flickers continually in and out when the Monitor Lizard is searching for food. Two species of crocodile (Malay: *buaya*) occur in Malaysia: the Estuarine Crocodile, which has a rather broad snout and feeds on a wide range of animals, and the Malaysian Gharial, which has a long slender snout, and feeds mainly on fish. The Estuarine Crocodile is more widespread, and may be found in freshwater rivers and lakes far inland, right down to the mangrove and coastal swamps. It is not a common animal, however, because decades of deliberate hunting have reduced numbers and made the survivors very shy. Loss of secluded nesting sites through forest clearance, building construction and aquaculture has also been instrumental in keeping crocodile numbers down. The largest surviving populations appear to be in the lower reaches of the broad rivers of Sabah and of Sarawak, where people are still very occasionally eaten by large crocodiles. Looked at in numbers of deaths per year, however, it is still much safer to bathe in a Sarawak river than to cross a road in any large city.

Turtles

Four, possibly five, species of marine turtle come to Malaysia's beaches to lay their spherical, white, soft-shelled eggs in the sand. Best-known, largest and most in danger of extinction is the massive Leatherback (or Leathery) Turtle. In Malaysia, it nests only within about twenty kilometres (12½ miles) of beach in the Rantau Abang area of Terengganu. Leatherbacks which have been marked with a tag in Terengganu have been relocated as far away as the seas off Hawaii, Taiwan, Japan, Indonesia and (mostly) the Philippines. The Leatherback egg-laying season

Green Turtles (*Chelonia mydas*) mating in the sea off Sipadan Island near the Sabah–Indonesia border. Turtles spend much of their time under water but, like other reptiles, are air-breathers.

lasts from April to September, but most of the turtles come up to the beach in June or July. The smaller Green Turtle is the most common species in Malaysian waters. It nests mainly on beaches near the Pahang–Terengganu border, on Perhentian and Redang islands, in south-western Sarawak and north of Bintulu, and on the islands off eastern Sabah. Most nesting is between January and October, with a peak between May and July. The Olive or Pacific Ridley Turtle nests on beaches in the northern half of Terengganu and in Sarawak. Most nesting is between January and October, with a peak between April and June. The attractive Hawksbill Turtle seems to be more abundant in the Philippines than in Malaysian waters, although it is doubtful that this state of affairs will continue in view of the numbers of immature Hawksbills caught annually there to be stuffed and sold as decorations. In Malaysia, Hawksbills nest on Redang island, Tanjung Geliga, the Pahang–Johor border and some islands off eastern Sabah. The main season is January to September, with a peak between April and May. The Loggerhead Turtle has yet to be positively confirmed as nesting in Malaysia, but it is possibly this turtle which has been reported laying eggs on the west coast of Sabah in the early months of the year.

Frogs and Toads

There are around one hundred species of frogs and toads in Malaysia. Some live only in deep, undisturbed forests, while others are widespread, and a few occur only in cultivated areas, like rice fields. They range in size from the tiny *Microhyla* frogs, which are only one-and-a-half centimetres (⅝ inch) long, and which live in low, damp spots in the forest, to the toad *Bufo asper*, which may exceed twenty centimetres (8 inches). This warty denizen of forested stream-sides is regarded as a delicacy by many rural people in Sarawak and Sabah. The so-called horned toad, actually a frog, has eyelids protruding out like horns. When sitting motionless on the forest floor, it resembles a pile of dead leaves. Wallace's Flying Frog has large, strongly-webbed feet which enable it to make long, gliding leaps between low trees and bushes. Some frogs spend their entire life in tree holes containing rain water, while others depend on the muddy wallows made by wild pigs and rhinos.

Freshwater Fish

Freshwater fishes abound in the less disturbed rivers and streams. Rural people everywhere regard the larger species as an important source of food, and various kinds of fishing methods, including nets, traps, lines and the roots of the poisonous woody climbing plant *Derris* are used. Unlike hunting, few methods of freshwater fishing are controlled by law. The larger fish are most popular, of course, with the *tapah* (*Wallago maculatus*) and the *patin* (*Pangasius* species) highly prized. Both occur in the larger rivers and their tributaries. The former may exceed one metre in length and weigh twenty kilograms (45 pounds). In contrast, the little *Protomyzon* fishes, about four centimetres (1½ inches) long, are often overlooked unless we search the bottom of the fast-flowing, clear rocky streams which are their home. Generally, small inland rivers and mountain streams retain their natural fish fauna. Unfortunately, the lower reaches of so many rivers nowadays are filled with various pollutants, that fears are being expressed over the survival of some species in Peninsular Malaysia. The Sarawak and Sabah fish fauna are now under threat in populated areas, too, because traditional fishing methods are being replaced by the use of highly toxic pesticides, as well as by portable generators, used to stun the fish which then float to the surface. Much of our knowledge of Malaysia's freshwater fish is anecdotal and we know less about their distribution and habits than we do of other vertebrate animals. Studies are particularly needed on their breeding requirements and tolerance of human activities, so that they can be managed as a renewable resource.

Life Amongst the Coral Reefs

Malaysia's marine waters are at the heart of a region known to biologists as the Indo-Pacific faunistic region. This means that marine life in the Indian and Pacific Oceans, and in intervening seas, shares close similarities and many species in common. Together with adjacent Indonesia and the Philippines, marine life achieves its greatest diversity of all in Malaysia.

The forms of life in a coral reef are stunning in their variety, shapes and colours. The living part of a coral is called a polyp. The hard part of the coral is made by countless polyps removing dissolved minerals from the sea water and depositing them as a solid mass. Each polyp maintains contact with other polyps by strands of living material – hence the numerous minute holes visible in the hard part of the coral. During the day the polyps hide away, and the corals appear dull and dead, but at night they seem to blossom, as the polyps extend their tiny, colourful tentacles to sift the water for food particles. Associated with corals are various kinds of fish (including eels, snappers, fusiliers, seabass, parrotfish, rabbitfish, triggerfish and angelfish), sea slugs, molluscs, sea squirts, sea anemones, sponges, worms, spiny lobsters and algae. All these forms of life are dependent on the corals. Some fish feed on coral polyps, nibbling off pieces with their powerful jaws, while others pull small invertebrate animals from crevices in the coral. Some specialize in eating sponges, others prey on smaller fish. The famous barracuda (*Sphyraena barracuda*) is a top carnivore, roaming the sea and reefs to feed upon smaller fish which themselves feed on tiny reef fish.

A Walk Through the Rainforest

When we look at a forest, we see its architecture: the trees, epiphytes and lianas, herbs, streams, rocks, the occasional bird or other animal. The source of energy that permits growth and movement is, of course, the sun. But the means by which the sun's energy is harnessed and channelled and reorganized to produce all this architecture is mostly hidden. It all starts with what we see everywhere in the forest: leaves. The sun's energy is captured by the leaves of plants and used to convert water and carbon dioxide from the air into carbohydrates. Other chemical elements may be added to synthesize other substances, ranging from enzymes to lignin, the basic stuff of wood. This is the beginning of a great web of interactions between numerous forms of life. We have to look very carefully, and exercise a fair amount of imagination, too, in order to understand that the forest teems with millions of tiny life forms, ranging from bacteria to bees, all of which are essential to its smooth running. The largest of these organisms (an organism is anything that lives and can reproduce itself) are invertebrate animals such as insects and worms, and some kinds of fungi. But most of the essential participants are microscopically small. And then there is a whole range of wildlife, from beetles to hornbills, which act to maintain the fine tuning of the forest.

In this section we shall be concerned, therefore, with the forest at work. We have already surveyed the larger animals which make their home here. It is now time to turn our attention to the smaller organisms, the beetles, termites, cicadas, butterflies and ants which process the stuff of the forest, not forgetting the equally industrious fungi. Along with soil and plants, they all function together as a system – the ecosystem.

Let us imagine a walk through the forest and try to uncover some of those many forms of life which remain hidden from the casual observer. The best time to see and hear a variety of wildlife, especially the larger animals, is early in the morning, starting at dawn. After a long, cool, dark night, many animals take advantage of the sudden burst of light and warmth of the sun's rays to seek food. The other good time of day is starting in the late afternoon, say about five-thirty, and continuing until well after dark.

The Early Morning Walk

Dawn is undoubtedly the freshest and most exciting time of day to be in the forest. At low altitudes, a wide, clear trail on flat land is the best choice for a walk, so that exertion is minimal, and it is possible to concentrate on looking at plants and animals, rather than on where our feet are going and what we are walking into. In the hills, cooler temperatures can make for a more pleasant walk, as long as we stick to ridge-top trails wherever possible, to minimize energy expended on going up and down, and on balancing upon steep slopes. Take a large container of water,

OPPOSITE PAGE A misty early morning in the rainforest – the Danum Valley in Sabah.

A party of walkers heading for Bario in northern Sarawak.

preferably with a little salt added (this will not be regretted several hours later), a compass (even experts can lose their way), and, in remoter areas, a *parang* (long knife) to cut down any vegetation that has grown across the trail, and to poke around in the leaf litter and rotting logs.

Concentrate on looking for things at a distance. Look up and around often. Look for movements and for colours other than green. Don't be too distracted by all the saplings and small trees nearest to you. Look beyond them. Listen: many things can first be located by sound, only later becoming visible. Don't be fooled by large dead leaves falling noisily through the tree canopy to the ground.

Gibbons and their Beetle Attendants

An unmistakable early-morning sound in the forest is the singing of gibbons. These delicately-built, slender apes will probably have started up before dawn – campers in the forest may have been woken by the plaintive solo song of the male. Later, the female of the group will give vent to long, loud calls which carry far into the forest, up to several kilometres across broad valleys. In Peninsular Malaysia, this consists of a series of high-pitched rising and then falling notes, while in Sarawak and Sabah, the female of a different species of gibbon instead produces a high-pitched, fast bubbling cry. If these sounds seem to come from nearby, it is worth trying to move in closer for a view of the gibbons themselves. While calling, they are less likely to flee from human observers. Normally, you will see a group consisting of mother, father and one or two youngsters, the smaller grasping on to one or other of the parents. Having finished their concert, the group will move away through the tree-tops with astonishing speed and grace, the spindly bodies and legs dangling like pendulum weights at the end of long, string-like arms. The gibbons' whooping cry should not be confused with another familiar forest sound, that of the Helmeted Hornbill. This starts with a single loud deep 'poop', which is repeated and then speeds up until it culminates in maniacal, braying laughter. Like gibbons, hornbills are fruit eaters and seed scatterers.

LEFT The White-handed Gibbon, *Hylobates lar*.

OPPOSITE PAGE Termites (here of the Hospitalitermes group) industriously construct nests out of soil and chewed wood. Columns of many thousands of individuals may often be seen snaking their way through the forest, only to disappear from sight in the trees or underground.

Gibbons are skilled at seeking the ripest of fruits from widely-dispersed forest trees and lianas. They eat the sweet flesh from the fruits and spit out or excrete the seeds intact. No doubt many kinds of forest plants which hide their fruits in a tangle of lianas or small, bushy trees, away from the eyes of larger animals, owe their continued existence to the feeding habits of gibbons. If you are of an enquiring frame of mind, you might like to stop and take a look at the droppings; there you may see the beginning of a new forest tree, a seed deposited with a ration of organic fertilizer. Be patient. Keep still and quiet, and you will be rewarded with two insights into forest life. Firstly, the sudden disappearance of the noisy apes will give the impression of a void waiting to be filled. You can imagine yourself looking into the forest from far away, as if your physical presence has not yet disturbed the pristine scenery. Gradually, new forms of life will come into view: small birds, insects or lizards. Watch them as they go about their business oblivious of your presence, and experience the forest as if no human being were there. Now look again at the remains of the gibbons' early-morning toilet and you may see a glimpse of the detailed workings of the forest: one or more blackish-coloured beetles, three to five centimetres (1–2 inches) long, will be industriously rolling balls of droppings through the leaf litter, eventually to disappear under the ground. What are they doing? They are taking home food for their young, which obtain their nourishment from the remains of partly-digested plant matter. The beetles not only clear the forest floor of animal waste, but channel it into the topsoil, along with any seeds which may be there, aerating and fertilizing as they go.

Squirrels as Consumers of Bark

Examine the varying textures and forms of bark on the different trees. You will be amazed at the variety, and at the tiny creatures – spiders, mites, ants and bugs – which inhabit these miniature landscapes. Looking up into the trees, your attention may be drawn to the seemingly frantic movements of small squirrels running along branches and up tree trunks, seeking food in the form of small insects, lichens and even the inner sappy bark of the trees. Some trees appear to have pock marks, or numerous tiny circular depressions in the bark, all over the trunk and branches. Look closely at the round depressions and you will see that some have been made quite recently and the marks of tiny teeth are still visible. These are the places where small squirrels have gnawed to feed on the nutritious inner bark. More curious is the fact that only a few tree species are favoured by the squirrels, and the various species are not closely related to one another. The sap of some types of *rengas* trees (family *Anacardiaceae*, the mango family) is sought by squirrels in this way, yet that very same sap causes severe allergic reactions in many people.

Longhorn Beetles: drillers of wood

If you watch squirrels gnawing away high above the ground, and see or hear fragments of dead wood showering down, they are feeding not on the bark, but either on the grubs of beetles, or on termites which have made their nest inside the wood. Beetles

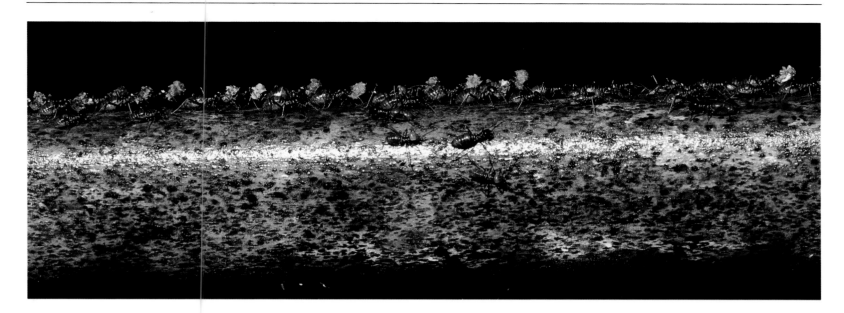

which lay their eggs into the bark of living trees are actually one of the most numerous forms of tropical forest life. Thousands of species are known. They are not very obvious because they tend to stay above ground level and are mostly the colour of wood. These wood-boring beetles are often known as Longhorn Beetles because of their very long, flexible antennae. From eggs laid into the trees, little grubs hatch out, which bore through the timber, growing larger as they proceed. The growth stage may take years, but eventually the adults emerge, leaving patches of dead wood or, in serious cases, a dead or dying tree. While these beetles are insidiously destructive, they probably help to maintain the great diversity of tree species in the Malaysian forests, because the more common the tree, the more likely it is to be found and injected with Longhorn Beetle eggs. Trees damaged by the beetle grubs are likely later to become infested with termites or fungi.

Termites as Consumers of Wood and Leaves

Termites are often called white ants; they are not true ants, but members of an insect group called *Isoptera*, to which cockroaches are related. They live in colonies of tens of thousands of individuals based in nests which, according to species, may be under the ground, on top of the ground, attached to tree trunks or actually inside the wood. Over one hundred termite species are known from Sabah alone, and each square metre of Malaysian dipterocarp forest contains literally hundreds of individuals, but normally they are concealed in their nests. Some species of termite have evolved a cunning means of utilizing lignin, the basic stuff of wood. They cultivate tiny fungi inside their nests (if a nest is opened, the fungi can be seen as white dots) and these fungi produce an enzyme that digests lignin. The termites eat wood, absorbing some nutrients but excreting the lignin. The fungi actually grow on the lignin-rich termite droppings, and the termites then eat the fungi. The fungi also produce an enzyme which can digest cellulose, the main constituent of leaves. By eating the fungi, the termites take in this enzyme, enabling them to digest leaves. Scientists at Pasoh in the lowlands of Peninsular Malaysia have estimated that nearly one third of all leaves which fall from trees there are taken as food by termites. The remainder are taken by bacteria, fungi, earthworms, millipedes, snails and other small creatures. In the mountain ranges, beetle larvae tend to replace termites as the main agent for breaking down and recycling wood and leaves into the forest system.

Bees as Utilizers of Resin

Occasionally, we may notice protruding from a crack in a tree trunk something resembling a small, pale grey-brown trumpet. This is the entrance to the nest of small, dark-coloured bees of the genus *Trigona*. The trumpet-like passageway is made from a mixture of tree resins and the bees' own wax. Small blobs of pure resin may be deposited at the entrance to gum up the legs of predatory ants, which would otherwise enter and feast on the bee colony. These bees do not possess a sting, unlike some Malaysian wasps, which not only have a vicious sting but chase after humans unlucky enough to stumble on their nests. Later on in the day, if we stop for a rest in the forest after a morning's walk, these same bees will find us and alight to lick sweat from our skin, and for this reason they are sometimes called sweat bees. Sometimes, instead of the sweat bees' trumpet-like nest entrance, we may see great splintery gashes on the side of a tree trunk, as if some frenzied lumberjack had scaled the tree and attacked it with a blunt axe. Once, the nest of a colony of sweat bees would have been here. The Malaysian Sun Bear, thwarted in any attempts to obtain the nests of the wild honey bees high up in massive, smooth-barked tualang trees, has to make do with the less tasty but accessible sweat bee nests.

Cicadas: consumers of sap

By mid-day, there is little obvious activity in the forest. In fact, the trees are working furiously but silently at capturing sunlight, absorbing carbon dioxide from the air and water and minerals from the soil, to grow and to produce flowers and fruits. The only sounds are from cicadas and from barbets, those shy, colourful birds which spend an inordinate amount of their time perched in the tree-tops making chonking sounds, not to be confused with the rattling and wailing of the cicadas. A regular call throughout the day is the repetitive 'chook-chook-chrrrrr' of the Yellow-crowned Barbet, gaudy but difficult to see as it sits motionless in the foliage – usually at a good height. Muller's Barbet, in the hill forests, has a rolling call 'tukaroo'. All the barbets, which are relatives of the woodpeckers, are basically bright green birds, but have brightly-coloured heads with patches of red, yellow or blue.

Cicadas are large insects which we rarely see but hear almost constantly in the Malaysian forests. Their sounds are made not from the mouth, nor, like many other insects, from rubbing two parts of the hard, segmented body together, but come instead

A cicada, a large sap-sucking insect, the most commonly heard and ubiquitous form of life in the lowland rainforests, but rarely seen.

Cut stems of certain liana's (*Uncaria* species) provide a rather astringent liquid, refreshing to the thirsty forest traveller.

from the vibration of a membrane across cavities in the sides of the body. The cicadas that call during the middle of the day have a distant, drowsy sound which from time to time increases in volume to become a loud high-pitched rattling. Usually, we cannot see the cicadas because, although numerous, they are right up in the tree canopy. Occasionally, we may disturb a solitary cicada which has landed low down on a tree trunk, but if we approach, it always manages to detect our movement before we can reach out to touch. It flies off as if bemused, bumping into branches and attempting to make its strident sound, but always failing. Possibly, these solitary, seemingly inebriate cicadas are those which have recently emerged from their long period underground as immature nymphs. Most cicadas lay their eggs into the stems or twigs of plants. The so-called nymphs which hatch from these eggs descend and burrow into the ground to spend a long period of growth feeding on the sap of roots. Some species spend years developing under the ground, progressing through several stages of increasing size. Then, suddenly, one day the final immature form emerges, pushing up a chimney of soil through the leaf litter, and settles on a hard support not too far above the ground. The adult bursts from the skin, which is left behind like some solidified form of insect ghost until rain and fungus cause it to disintegrate.

The Mid-day Lull: butterflies and lizards

If you have already drained your water bottle and are still thirsty, you could seek out a liana of the genus known as *Uncaria*. By cutting through the thick woody stem you will release a flow of water which is quite palatable, if somewhat astringent to the taste. The hairy rhinoceros is also fond of its leaves and twigs, seeming to prefer the bitter flavour. At mid-day we can see very few signs of insect activity, except for butterflies, which may be abundant during the middle months of the year, but impossible to find at other times. An unforgettable sight in the Malaysian

dipterocarp forest is that of the Common Tree Nymph butterfly (scientific name *Idea iasonia logani*), with its ghostly, dark-spotted translucent wings, as it alternately flutters and glides through the tree-tops. Equally memorable but totally different is the exotic flash of colour of the Dark Blue Jungle Glory butterfly (*Thaumantis klugius lucipor*), which is found only in the gloomy, thickest parts of the forest. As suddenly as it appears, it disappears, because when it settles, the dark colour of the undersides of the wings blends in with the tangle of dead leaves.

Relying partly on heat from the sun's rays to provide energy sufficient for movement, small forest lizards catch our attention as the day warms up. Burnished coppery skinks skitter through the carpet of dry leaves on the sides of the trail, waiting for insect food to alight, but distracted by our movement and in turn distracting our attention as the leaves scatter and crackle. Look out especially for tiny gliding lizards (*Draco*), usually just above the level of our heads. Whether sitting or gliding, they are something of a novelty for the newcomer to the forest. A quick movement between two tree trunks, spotted out of the corner of our eye, is often the first clue we have of this little creature's presence. It is able to glide (but not fly) because some of its rib bones extend well out from the sides of the body and between these ribs is a skin membrane, which is folded down when not in use for gliding. The lizard may stand motionless, head upwards on a tree trunk, except for a flag-like flap of brightly coloured skin on the throat which is repeatedly opened and closed.

Often in the rainforest a long period of apparent inactivity ends as a bird party, or 'bird wave', moves through – a hunting association of many different species, whose members seem to derive mutual benefit from their various foraging activities. Leafbirds, or other nectar feeders like Whiteyes and Sunbirds, disturb insects which are quickly snapped up by Flycatchers as the informal band progresses, scouring the forest at every level from the canopy down to the ground. So rapid is their passage that it is often difficult to identify all the species participating.

Gliding lizards (*Draco* species) spend most of their life on tree trunks but lay their eggs in a small hole in the ground, as shown here.

Ants, in terms of numbers of individuals probably the most abundant creatures of the lowland rainforests, keep the forest clean. Here, a small group, working in selfless co-operation, drag the carcass of a dead scorpion back to their nest to serve as food.

A Walk at Dusk

Apart from these occasional flashes of life, most animals become sleepy and lapse into inactivity after mid-day. We should follow their example until late in the afternoon. Then, in preparing for an evening walk, we can dispense with the water bottle and instead make sure to bring a torch or flashlamp with new batteries and a spare bulb. Start the forest walk well before dusk, so that you have an hour or so to look around before it becomes necessary to use the lamp.

It is late in the afternoon and after dark that many kinds of ant become most active, appearing as if from nowhere, but actually from underground nests and crevices between roots. Most fearsome looking because of their size are the reddish-brown and black giant ants (*Camponotus gigas*) which start wandering along forest trails in dribs and drabs in the afternoon, but are harmless, almost friendly to people. Only one type of ant, the strictly nocturnal fire ant, causes us major problems in the forest, and then only in camp where thousands may swarm over everything in sight. To say that each bite feels like the stab of a red-hot needle is no exaggeration. While beetles and termites are amongst the most common form of insect life in the Malaysian lowland forests, the ants are truly the most abundant of all. At Pasoh in Peninsular Malaysia, an incredible average of 2,000 ants inhabit every square metre of soil. Ants are the main predators of the insect world in the lowlands, but up in the mountains, centipedes and spiders are more common. There is a fair chance of seeing scurrying centipedes or crawling millipedes on the forest floor. The large, reddish-brown centipedes may give a very painful bite, while the even bigger, but vegetarian, shiny black millipedes are harmless except for the ability to give off a stinking protective liquid suggestive of a school chemistry laboratory.

As the sun starts to set, darkness quickly falls inside the forest. Now, one of the most memorable sounds of the Malaysian forest

Numerous forms of invertebrate life help to maintain a constant recycling of nutrients in the rainforest system. Here a mass of centipedes have been exposed by removing leaf litter from the forest floor.

may be heard – a sort of wailing trumpet sound repeated over and over again, as the so-called 'six-o'clock cicada' has its say. (In Peninsular Malaysia, where dusk falls later than in Sarawak or Sabah, this cicada makes its noise at seven o'clock.) So powerful is the sound that many people, when asked to guess, believe it to be the call of a hornbill or other large bird. You may be startled at this time of day to hear what one might guess to be a demented soccer referee suddenly blowing his whistle as loudly as possible into your ear. This, of course, is another cicada, and the remedy is

ABOVE A small group of the estimated two million or so bats which inhabit Gomantong Caves in Sabah. They emerge around dusk, wheeling through the sky as they disperse in search of food.

BELOW Spiders abound in the rainforest, from the forest floor up into the trees. Many – even the large ones – are well camouflaged and seen only at dusk, when a torch may pick them out by the reflection of light from their tiny eyes.

to approach the offender, which will fly off, immediately reducing its racket to a gurgle. Just before complete darkness approaches, a further cicada starts its song, a gentler, lazy sort of humming sound that is so appropriate to end the daylight hours.

If you happen to be in the vicinity of one of Malaysia's great limestone cave systems, this is the time of day to witness one of the world's most astonishing spectacles. Thousands upon thousands of bats sweep out of the caves, wheeling and spiralling in the air, on their nightly search for food. They will fly in formation for up to fifty kilometres (30 miles) before dispersing to snap up the insects which they have detected with their sophisticated echolocation system. Look out also for the hawks which station themselves around the caves to swoop and catch both exiting bats and entering swiftlets.

Only after nightfall do all cicada noises cease and the curt whistles and plinks of crickets, nocturnal members of the grasshopper family, and small frogs take over. With our flash-lamp on, the forest seems to surround us very closely, for so much light is reflected off the nearest saplings, bushes, tree trunks and lianas. Little light penetrates more than a few metres away. The motionless pinprick eyeshine of spiders – all harmless in Malaysia – and of small frogs distract attention. In forests near to swamps or rivers, we will be beset by hordes of mosquitoes, but the higher we go above sea level or the further away from a river, the fewer they will be. Rarely are glimpses caught of nocturnal mammals such as one of the civet family, or the Moonrat (the world's largest member of the insectivore family). The eyes of almost all nocturnal Malaysian mammals reflect torchlight at night, and this is the best way to detect their presence. (The only exception is the Tarsier, which is absent from Peninsular Malaysia.) The Moonrat, though, with its rather small, unalert eyes is often first spotted by sight or smell. In northern Borneo, normal Moonrats are albino. It is quite an experience to see this white furry denizen of the forest floor shuffling along damp depressions in the ground, searching for earthworms and giving off a distinctive rank, sweaty odour.

Another smelly nocturnal mammal, which in Malaysia is confined to the Borneo states, and which is most commonly detected on the forest edge, is a member of the badger family, known locally as the *tudu* or *teledu*. The odour is much like that of the North American skunk, vaguely reminding one of burning rubber. Most rural people claim that the smell is due to the animal breaking wind. But this badger possesses an anal gland, the secretion of which it can squirt out to a distance of a couple of metres. It is as well to avoid the firing line of this mucus-like malodorous blob.

It is a good idea occasionally to turn your flashlamp off for a few minutes. This will heighten your sense of hearing: all the forest sounds become very clear, and even the munching and marching of termites and ants may be heard. Little spots of yellow light may be seen, either moving very slowly through the leaf-litter or through the air, flashing on and off. Both lights are produced by fireflies, which are actually a type of beetle, the former by the adult females and larvae, and the latter by the adult males. The light is produced at the end of the abdomen, by oxidation of a fatty substance known as luciferin. As our eyes become accustomed to the darkness, something even odder may become apparent: faint luminous white patches, actually a kind of fungus, cover fallen branches, twigs and leaves. In some places, after a rainy period, delicate mushroom-like fungi of the genus *Mycena* appear on rotting logs or bamboo. In the darkness we can see that these, too, are luminescent, but green or blue, and much brighter than the white encrusting fungi. An hour or two after dusk, most will have gone, eaten by beetles which are presumed to be attracted by the light and which act as dispersers of the fungal spores.

The Work of Fungi

Fungi are much more than mere curiosities of the forest. The common feature of all kinds of fungi is that, unlike true plants, they are incapable of making any of their own foodstuffs. Everything has to be absorbed from true plants, either dead or alive, or from organic constituents of the soil. Of all the forms of life which we have met so far, fungi are perhaps the most essential to the functioning of the Malaysian rainforests, because they alone are able to cause the complete chemical breakdown of wood, thereby releasing nutrients which, directly or indirectly, end up supporting all manner of life. Short-lived mushrooms and toadstools can be seen everywhere on dead and dying wood after rainy periods, when they grow at a tremendous rate to produce millions of spores. The similar but hard and longer-lasting bracket fungi are even more prominent. But much less apparent are all manner of tiny fungi, called mildews, rusts and smuts which infect healthy leaves and stems, gradually draining life away through minute thread-like passages called *hyphae*. These kinds of fungi are a nightmare for farmers and gardeners, but in the forest they act, just like Longhorn Beetles, to maintain the diverse web of life, preventing any one kind of plant from coming to dominate the forest. And then, totally invisible to our eyes, is a vast array of microscopically small fungi living in the soil, and called *mycorrhizae*. It is only in recent years that scientists have come to be aware of the importance of these *mycorrhizae* to tropical forest trees. What apparently happens is that these minute fungi live in close association with the roots of trees, including the dipterocarps, taking some nutrition from the trees but in return helping to feed the tree with nutrients which are in very short supply in the soil. Trees infected with *mycorrhizae* will thus grow better than trees without them.

Tropical rainforests are the most complex ecosystems on earth. Only a small fraction of the many forms of life which are found in these forests have been identified, described and named by scientists. Most of those that have are plants and vertebrate animals. Recent investigations of life in the tree-tops suggest that the number of insects, for example, may be several times greater than previously thought. The science of 'rainforest biology' is indeed still in its infancy. The diversity of life forms is so bewildering and the complexity of their relationships is so great, that it is perhaps easier and more useful to think of them in terms of processes rather than as a list of species names. Looked at in this way, anyone can become an amateur rainforest biologist. Go out into the Malaysian rainforest, shed any preconceptions you may have about it, and study carefully anything that takes your fancy. Possibly you will be able to observe and perhaps even understand something which is completely new to science.

ABOVE Fungi are the major agents in the breaking-down of wood and other plant material into a condition usable by other life forms. This is possibly *Dictyonema*, a basidio lichen formed by a symbiotic union of a fungus and an alga.

BELOW A dead branch has fallen on to the forest floor, killed perhaps by the activities of termites or beetle larvae in the tree canopy above. The wood will now be converted into soil nutrients by the action of fungi and bacteria.

The Peoples of Malaysia

Human beings have lived in Malaysia for a very long time. Across the Straits of Malacca in northern Sumatra, an immense volcanic eruption occurred about 75,000 years ago. It was then that Lake Toba was formed and ash was spread over a wide area, including present-day Peninsular Malaysia. Stone-Age axes have been found below the level of the ash, indicating that people were already living there. The earliest inhabitants seem to have spread from mainland Asia southwards to Australia and it is possible that the Negrito Orang Asli of the hilly northern parts of Peninsular Malaysia are their descendants. Evidence of human activity from about 40,000 years ago has been found in the Niah caves in northern Sarawak and we may assume that other people were living elsewhere in the region at the same time. Excavations in the Niah caves have shown an extraordinary series of changes in human and animal life over tens of thousands of years. It is intriguing to find that large animals such as rhinoceroses, Orang-utans and Pangolins are now smaller in size. Possibly this has something to do with changes in climate and vegetation, or the availability of essential minerals. Remains of Tapirs, a species still present and widespread in Peninsular Malaysia, occur in the Niah caves until fairly recent times, yet they are now extinct in Borneo. Wild pigs were always a major source of meat, but a whole range of animals were taken as food, from snails to Orang-utans. The earliest clear signs of human activity are coarse chopping tools. By 12,000 years ago, more delicate, versatile, flaked stone tools appear, and it is likely that bone arrow-heads and spear-tips were made, increasing the range of species that could be obtained as well as the rate of success. By 4,000 years ago, rounded stone axes occur, presumably designed to cut down the forest in order to plant crops. By this time, too, pigs and dogs had been domesticated and pottery, nets and mats were being made. At 2,000 years ago, more elaborate pots and some locally-produced metal appear in the record and at about 1,300 years ago, imported iron, glass beads and ceramics are found. The production of hardwood blowpipes, for use with poisoned darts to kill animals, requires the use of iron tools.

Evidence from the Madai-Baturong cave region of south-eastern Sabah suggests that people did not remain in one particular place for long periods. More than 20,000 years ago, there was a spate of volcanic activity in south-eastern Sabah, and a volcanic lava flow blocked the Tingkayu river at one point, forming a lake. At some stage after the formation of the lake, people came and settled around its margins, leaving their stone tools to be discovered in recent years. About 3,000 years later, the lake drained away, exposing a cave in a limestone hill now called Baturong, and people lived in the caves there, at least from time to time, until about 12,000 years ago. Sea levels were then much lower than they are now. Borneo would have been joined by swamps and indirectly through Sumatra by dry land to mainland Asia. The clear blue waters of Darvel Bay, to the east of Baturong caves, were also dry land stretching to the present Philippines island of Sibutu, and the present array of islands were hill-tops dotted across a broad plain. Then, for unknown reasons, the cave-dwellers disappeared. From about 10,000 years ago, people occupied instead the bigger Madai cave system further east, until around 7,000 years ago, by which time the sea had risen to about its present level. Then, another gap occurs in the record, and Madai caves were devoid of human life until 3,000 years ago, when pottery first appears. Later, copper and iron

Penan headman and his wife, photographed in the Mulu district of Sarawak.

tools and beads from India were brought to the caves, but then yet again, around 500 AD, the site was abandoned. At some time later, probably after 1600 AD, people returned to the caves, not to live permanently, but to collect from the cave walls the edible nests of small birds called swiftlets. The nests, made of the birds' sticky saliva, were and still are exported to Chinese communities for the preparation of the delicacy known as birds' nest soup.

The earliest-known Chinese accounts of Peninsular Malaysia were written in the third century AD and at around the same time Indian traders reached the area. In the centuries that followed, explorers and traders from both China and India arrived, and their diaries give some idea of Malaysian life in those days. Different river systems were each ruled by a king: rice was grown and all the kinds of domesticated livestock that we now know were already being reared. Salt was obtained by boiling sea water. Alcoholic drinks were made from coconuts and rice, and perfumes were worn. All manner of natural products went out of Malaysia mainly in return for pots, porcelain, lacquer-ware, metal items, musical instruments and cloth. Major exports from the earliest times, and even up to the present, included the aromatic resins of forest trees, gahru wood, camphor wood and other hardwoods, ivory, rhino horn, hornbill casques, bees' wax, cowrie shells, tortoiseshell, rattan, coconuts, bananas, other edible fruits, betel nuts, pandan mats, bamboo, gold and tin. Much later, possibly not until the eighteenth century and especially from eastern Sabah, came the edible swiftlet nests.

By the seventh century AD, Kedah, at the north-west of the Peninsula, had evidently become a major centre of human population, culture and trade, largely through the influence of Hindu and then Buddhist settlers from the east coast of India. Several aspects of Malay culture, such as details of the ceremonies relating to births and weddings, date from early Indian influence. In the tenth century, Chinese and Arab traders were

Over the past thousand years, the great island of Borneo – being further from the trade routes of India, the Arab region and Europe and much larger in size and more mountainous than Peninsular Malaysia – experienced less influence from traders and immigrants. Away from the coast, different tribes can be recognized by differences in language and culture. Before this century, the differences were evidently taken very seriously by the people themselves, and there were frequent inter-tribal wars, which both caused and were caused by long-distance overland migrations. In many aspects of life, the tribes were similar: most lived in multi-family longhouses, practised shifting cultivation, drank rice wine and believed that calls and sightings of certain birds and snakes were important omens. Several tribes shared such features as distinct social classes within a community, the practice of tattooing and of women wearing brass ornaments. Today, the Ibans form the largest indigenous group of East Malaysia. Their language is closer to Malay than any other indigenous language in Sarawak or Sabah, and the Ibans are believed to be one of the most recent migrant groups from Indonesian Borneo, yet few have adopted the Malay way of life.

Describing Malaysia's present-day population is a complicated business and any attempt to put them into categories is bound to be somewhat arbitrary. Here, we opt for a description which is different from that which you will find in most other books and one which is more in tune with our theme of wild Malaysia. We shall classify people according to their environment, rather than by culture or race, dividing them into forest nomads, hill forest rice growers, riverside people and coastal fishermen and traders. Urban and plantation communities are excluded because similar groupings can be found elsewhere.

Forest Nomads

Nomadic, forest-dwelling people who subsist exclusively or mainly on native wild plants and animals still live in the hilly northern regions of Peninsular Malaysia, where they are known as Negritos, and in northern Sarawak, where they are called the Penans. Negritos can be encountered in and around Taman Negara, while Penan people live in and around the region of Gunung Mulu National Park. The Negritos traditionally obtained their carbohydrate diet from the tubers of climbing plants of the genus *Dioscorea*, but the Penan derive it from the stems of a palm of restricted distribution, *Eugeissona utilis*. Both obtain their protein from the meat of various wild animals, predominantly monkeys despatched with blowpipes and poisoned darts, and wild pigs killed with spears. These people possess the closest remaining links with the rainforests of Southeast Asia. They it is who can tell the identity of a plant from one leaf better than any botanist, and the identity of an animal from its track or sound better than any zoologist; they know best the properties and uses of all manner of forest produce and live in closest harmony with the natural environment. It is sad, therefore, that modern technological progress has created a dilemma both for these people and for the Malaysian government. The government is, of course, rightly obliged to provide forest people with the opportunities enjoyed by all other communities in Malaysia, but most of those opportunities are incompatible with a continuing forest life. Forest nomads have to make a basic choice between two life-styles, and there is the crux of the problem. Few individuals can make the switch – communities have to absorb change gradually. We can only hope that change follows a slow and gentle course, with exchange of knowledge between nomadic and settled communities, so that new opportunities can be grasped without totally losing the old ways.

Penan children at a settlement in the Mulu district.

using Kedah as a meeting place. From the seventh to thirteenth centuries, Kedah was a part of the Sumatra-based Srivijaya empire but that empire crumbled and the northern part of Peninsular Malaysia came under the greater influence of Siam. In the south, the Java-based Hindu Majapahit kingdom took Singapore, then called Temasik. Islam seems to have been introduced to Malaysia by Muslims from the west coast of India, some time after the tenth century. Adoption of Islam by coastal communities throughout the region came during the following centuries and by about 1400 AD the Idahan people of the Madai area in Sabah were already Moslems. In 1403, a Malay king from Palembang in Sumatra, married to a Majapahit princess, settled and founded the port of Melaka, which grew to become the major kingdom of Peninsular Malaysia, trading in goods from China, India, Sumatra, Java and the east Indonesian islands. But in 1511, the Portuguese conquered Melaka, followed by the Dutch in 1641, who ceded it to Britain in 1824. Although Indian and Chinese migrants had settled in Malaysia since the earliest times, it was the British rule that encouraged large-scale immigration of Chinese and actively brought in Tamil Indians to work on rubber plantations.

Hill Forest Rice Growers

The first rice plants introduced into Malaysia were probably not 'wet rice', which is planted in water-filled fields, but 'dry rice' or hill rice, which is grown on dry land, although not necessarily on slopes. Six months of work, starting with the felling and burning of the forest and ending with the pounding and winnowing of rice grains after harvest can supply a family with most of its food requirements for the following year. Growing rices makes for a more stable and less precarious existence than leading a nomadic life, but cultivating dry rice brings three problems to contend with. Firstly, rice grains are the food of several species of small birds known as munias – they are called 'pipit' in Malaysia, but are not related to the pipits of northern climates. Visits by munias to a ripening rice field can wipe out the whole crop unless it is guarded continuously throughout the day. Secondly, the soil loses its fertility with repeated planting, and, because fertilizers are not used, new areas have to be opened up. Thirdly, weeds spring up and grow vigorously and continuously in the non-seasonal tropical climate, making cultivation a progressively arduous task. Taken together, these two latter problems mean that the only way to sustain rice production is to shift to new forested areas every one or two years. The system used by people dwelling in the hills is to clear patches of old secondary forest, allow the wood to dry, burn it, and plant rice grains into holes made in the ground with a shapened stick. After harvesting,

three or four months later, the field is abandoned and natural vegetation allowed to take over. Next year, a different patch of forest is cleared for rice production. This is termed shifting cultivation. Seen from a distance, it produces a patchwork of different stages of regrowth dotted over hillsides. After twenty years or so, a field will have become forest again, with trees of a few species all with light, easily-felled wood, and all of similar size. After eighty years or more, the forest will be tall, much richer in species, and resembling the original, natural forest again. The system is self-sustaining and needs no fertilizers. But it depends on small fields surrounded by forest, so that soil is not washed away into rivers by heavy rain, and so that seeds from surrounding trees, rather than of herbaceous weeds, can fill abandoned fields. And it depends on a density of less than ten people per square kilometre, so that at any one time, only a small area is under cultivation. Newly overgrown fields provide lush food for deer and wild cattle in a habitat otherwise poor for these animals. In the past they provided a major meat source for shifting cultivators, but the widespread use of shotguns has made wild animals rare and wary in regions inhabited by hill folk. The system is basically good, more ecologically sound and sustainable than many of the more sophisticated means of food production, but once too many people practise it in one area, the forest will not regrow. Instead, coarse grass called *lallang* takes over, growing so vigorously that neither tree seedlings nor rice can ever compete.

The patchwork of vegetation produced by the practice of shifting cultivation, a common sight in the hills of Borneo.

Living in the hills made sense in days gone by and still has some advantages even now. The climate is cooler than in the lowlands, more suited to hard physical work in the open. Water-borne and mosquito-borne diseases can decimate lowland riverside communities, but this is rare in settlements around rushing hillside streams. There are fewer pests, both in numbers and kinds, on rice and other crops in the hills. Amongst the peoples who adopted this way of life are the Temiar Orang Asli in Peninsular Malaysia, most of the inland peoples of Sarawak, and the Muruts and Dusuns of Sabah. Lack of contact with foreign traders and missionaries has meant that few hill-dwellers have adopted Islam or Christianity as their religion.

Riverside People

Life in the hills also has disadvantages: poor communications with other communities, insufficient commodities such as metals and salt, genetic inbreeding, malnutrition in years of bad weather and poor harvest. Many of these problems are alleviated by living on the banks of rivers navigable by boat. Here, too, the alluvial soils are fertile and fish can be caught. Rice fields are still shifted from time to time because of the weed problem. This was the way of life of many Malay communities in the past and is still practised along the rivers of Sarawak and Sabah. In Sabah, these people are actually called Orang Sungai (river people). The main disadvantage of life on the riverbank is the risk of flooding. Major floods occur unpredictably every few years, damaging crops and houses, changing the course of rivers along their lower reaches, occasionally washing away entire villages. Orang Sungai on the lower Labuk river, which has as its source the Liwagu stream near to Kinabalu Park headquarters, used to attribute major floods at the beginning of the year to the birth of a dragon in a lake (now the town of Ranau) from millions of tiny hatchling fish, known as *mentigor*, which could be seen migrating up-river a few months previously.

The maintenance of weed-free fields is the most time-consuming activity when growing tropical crops, and it was probably for that reason that wet rice, or *padi sawa*, was developed. Few weeds can compete with rice seedlings which have been planted in open water. Wet rice can be grown wherever the soil is fertile, where there is adequate water and where the ground is flat or can be made flat. Extensive coastal flat-land not subject to excessive flooding is an obvious choice for wet rice cultivation, and we can see how much land has been developed in many parts of Malaysia, but especially on the north-west coast of the Peninsula. Wet rice was introduced into South-east Asia by Hindu Indians (as were the plough, cotton and the spinning-wheel) and it is likely that Kedah was the first major site of wet rice cultivation in Malaysia. Fertile soils in the hills can be made into terraces and irrigated by diversion of streams for growing wet rice. Large and successful communities have developed in this way in the upland valleys of Ranau, Tambunan and Bario, all in north-western Borneo.

Fishermen and Traders

Communities on the coast, whether along sandy beaches or mangrove swamps or at river mouths, often have inadequate land available for rice cultivation and instead rely largely on the harvesting of fish and other marine produce for food and trade. Many centuries ago, the right combination of resourcefulness in certain individuals or communities, and lucky contacts with traders led to the blossoming of some coastal communities into regional trading centres and, in a few cases, sultanates with wide

spheres of influence. These are the bases of Malay culture, of the configuration of the present Malaysian states and of the Malay language, which evolved to serve as the common means of communication throughout the entire region. Malay people include many races from all parts of South-east Asia, including what are now Malaysia, Indonesia, Brunei, the Philippines and even Thailand and China. The only features which all present-day Malays share are their religion, Islam, and a fluency in the Malay language. Movements of Malay peoples across the seas has continued for centuries. The ancestors of many Malays in Peninsular Malaysia came from Sumatra, Java or Sulawesi in Indonesia, while the Bajaus of Sabah, who are Malays according to the customary definition, may have come from Johor or the Philippines and possibly both places.

Finally, what of the 'wild men' of Borneo? Did they and do they exist? As an individual, one of the wildest men to have ever lived in Borneo was an Englishman named Raffles Flint who in 1890 boated and walked his way with a party of twelve Iban policemen across what was then one of the remotest parts of Sabah, to slaughter more than 130 natives, one of whom he suspected of killing and cutting the head from his brother. Headhunting – that is, killing people, cutting their heads off and preserving the skulls – was indeed one of the practices of many of the native peoples of Sarawak and Sabah during Flint's time and the activity was widespread in Borneo earlier in the nineteenth century. Shocked European settlers, who condoned the imprisonment of natives for failing to obey new laws which they could not understand, and the hanging of those who fought back, felt it their duty to stop headhunting, and with rare exceptions they proved successful within a few decades of rule. Perhaps the outrage felt by foreigners was in part due to their failure to understand why people wanted to cut the heads from others. These were the days before anthropologists and ethnologists went galivanting in the tropics to observe and ask questions, and it is indeed rather difficult nowadays to know how headhunting evolved and was maintained as a normal thing to do. It appears that for a few tribes, such as the Kenyahs and Kayans, religion demanded that human heads be obtained for use in ceremonies after the death of a leader. Apparently, other tribes in the region, adversely affected by this requirement, retaliated and long-term feuds evolved. Thereafter, headhunting became something of a sport, and also a good way for a young man to show his prowess to a young woman. For many Ibans, a particularly vigorous and mobile tribe in the eighteenth and nineteenth centuries, human heads became an important part of the dowry required of young men to obtain a bride. Knowing the apparent toughness yet innate gentleness of most Borneo people, it is easy to believe that giving up headhunting was no hardship for them and that this was one of the ways in which many willingly co-operated with the European settlers. Headhunter scare-mongering still surfaces every few years and at such times naughty children are threatened with a visit from the headhunter as the local equivalent of the bogeyman.

Another story which refuses completely to go away is that of the existence of short people, Orang Pendek, usually rather hairy, who live in remote forests and flee from us tall people. These can be explained away as sick old Orang-utans or monkeys, or where only footprints have been seen, as the rained-on tracks of bears or wild pigs. And yet, two highly experienced and reliable people in both Sarawak and Sabah have assured me that they have clearly seen Orang Pendek during their forest travels.

Wildlife Conservation

The bones of prehistoric animals, found at Niah and other caves, show that people have hunted animals for food for at least tens of thousands of years. The intensity of hunting has, if anything, increased in recent decades because of the increase in the human population and the use of guns. In some areas, this has contributed to the rarity and even extinction of species. Hardest hit have been the rhinos, whose horn has for hundreds of years been in demand from Chinese communities throughout Asia for supposed medicinal properties. The so-called Javan Rhino (now confirmed as present only on the island of Java in Indonesia) has disappeared entirely from Malaysia, while the hairy (Sumatran or Two-horned) rhino occurs now only in small, scattered populations. It may be that natural circumstances have partly led to the decline of rhinos, but there is no doubt that hunting has aided their path towards extinction. With the possible exception of rhinos, hunting may have contributed to the rarity of other large animals in certain areas but it seems not to be the basic cause of extinction, at least not in Malaysia.

The Problems of Translocation

The main reason why the number of large animals has been decreasing both in Malaysia and around the world is because so much forest has been converted to agriculture and other kinds of land usage. It is a common misconception that animals can move (or be moved) about at will and survive happily under new circumstances. A particular area can support only a particular number of animals of any given species, depending on the availability of essential resources such as food, water and shelter.

Take the elephant as an example. In the partially forested floodplain of a large river, with lots of open grassy areas and fertile soil encouraging quick growth, one hundred square kilometres can support perhaps ten elephants. Up in the forested hills, rarely is a blade of grass to be seen, and a hundred square kilometres probably supplies enough food for only one elephant. A group of ten elephants would require one thousand square kilometres in which to live. But then they are faced with a difficult problem: they would have to walk so far to obtain a little food, that they would run out of energy before they could make it to the next patch of grass.

The inference is that if forest is cleared in the lowlands for a plantation, animals do not run away to a new life in the hills. Instead they die or, at best, cease breeding. What if there is still some forest left in the lowlands? Animals have legs (or wings). Surely, they can move? Wrong. Let us say, for example, that ten Orang-utans lose their home when a new plantation replaces their forest. What happens is that some of them, especially the youngsters, totally unused to chainsaws, get caught as the trees come down. Lucky ones are taken by kind workers and sent to the Orang-utan Centre at Sepilok. Unlucky ones die from injuries or starvation or from not-so-kind workers. The older and more resourceful Orang-utans can move into adjacent forest, but there, more unpleasantness awaits them. That forest is already full to capacity with other, resident Orang-utans, who are not going to tolerate interlopers stealing their food and their mates. The poor interlopers, unfamiliar with their new surroundings, do not have the experience to seek out the best food trees. If they do find food, the chances are that they will be excluded from enjoying it by the original residents. The fact is that only a few

animals move into new areas when their habitat is destroyed, and those that make it are doomed to a life of stress, insufficient food and little hope of breeding.

Apart from a few special cases, attempts to transport wild animals from one place to another are doomed to failure. Wildlife departments, conservationists and others concerned with nature conservation and animal welfare are placed in a terrible dilemma. Projects to move elephants, for example, from isolated patches of forest into a park or reserve are widely publicized. The aim of such translocation projects, as they are called, is to save elephants otherwise doomed to die from starvation, or from the guns of irate plantation managers, or even from eating vegetation contaminated with pesticides or herbicides. If we ask ourselves some very searching questions, we may wonder what really happens to those animals which have been translocated, and if they are much better off. Perhaps conservationists must share some of the blame for being party to misguided attempts at kindness. And perhaps we should instead put all our efforts into demonstrating to the decision-makers that there is no substitute for a system of parks and reserves based on the needs of wildlife, not on what is left after the choicest bits of land have been taken away.

Some Malaysian Conservation Programmes

While there is no entirely satisfactory substitute to conserving wildlife in a natural habitat, it is sometimes helpful to provide special measures so as to boost the numbers of endangered species. Some important conservation programmes have already been touched upon in other sections of this book. The wildlife authorities of Peninsular Malaysia and Sabah have developed a scheme whereby Sumatran Rhinos, when found isolated in unprotected forest, are being brought into captivity to establish a managed breeding population. Sepilok Orang-utan Centre in eastern Sabah was founded in 1964. Since then young Orang-utans which have lost their mother, either through being captured deliberately, or through destruction of their forest home, have been brought to Sepilok so that they may live in a natural environment. So far, over two hundred Orang-utans have gone to Sepilok. Sadly, the death rate has been rather high – various diseases take their toll, including malaria and pneumonia. On the bright side, health care has improved in recent years, and several Orang-utans which have been brought to Sepilok as infants have reached maturity, mated with wild Orang-utans in the forest, and given birth to youngsters. There is no doubt that Sepilok has saved many Orang-utans which would otherwise have died, and at the same time the fact that the project is open to the public has meant that many children in Sabah have developed a love of nature through visits to this reserve.

Coastal people throughout Malaysia traditionally regard turtle eggs as a delicacy, and the eggs fetch high prices in the towns. These factors in combination have resulted in many years of intensive egg-collecting, which no doubt has increased in recent decades along with the increase in the human population. Modern egg-collectors go to the remotest of beaches during the laying season, and throughout Malaysia only a very tiny fraction of eggs laid every year are permitted to hatch. Since turtles are believed to require many years to reach maturity, and to live for a long time, the effects of too much egg-harvesting may take decades to become apparent. However, all turtle populations in the region now seem to have been on the decline for many years. The degree and kind of legal protection afforded to turtles, or to be more precise their eggs, differs between Peninsular Malaysia, Sarawak and Sabah. In recognition that some kind of control on egg-collecting would be needed to save sea turtles from eventual

Leatherback Turtle egg hatchery at Rantau Abang, Terengganu.

extinction, a system of collecting a proportion of newly-laid eggs and transferring them for protection to an artificial 'hatchery' was started for Leatherbacks in Terengganu in 1961. Since then, other such hatcheries have been started for other turtle species in Kelantan, Pahang and Perhentian island, Sarawak, and on the Turtle Islands off Sandakan in eastern Sabah. Despite these measures, turtle numbers are still on the decline. Progress achieved by the hatchery projects seems to be partly cancelled out by modern developments. Hundreds of mature turtles die accidentally every year in Malaysia, caught in trawling nets. Others eat and choke to death on some of the millions of plastic bags which have been dumped into the sea in recent decades, and which turtles mistake for jellyfish or other food items. Also, many nesting sites are permanently disturbed by new residential and industrial developments on the beaches. The Malaysian government is concerned about this trend and in co-operation with WWF Malaysia, proposals for a new, more comprehensive national turtle conservation programme are now being made. Let us hope that some headway can be made against the avalanche of undesirable side-effects which result from techno-logical progress.

The Need for Plant Conservation

Having discussed animal conservation at some length, it should be noted that too much emphasis has been placed, by all parties concerned with conservation, on animals – especially the large ones. This has been very much to the detriment of plants and invertebrate animals. During the past century, only two species of mammal have become extinct in Peninsular Malaysia, despite the massive loss of forest. These are the Javan Rhinoceros and the *banteng*, both of which were approaching extinction by the turn of this century, possibly as a result of natural causes. Probably no human efforts could have saved them. As far as I

know, no species of bird has become extinct, and amongst reptiles only the Leatherback Turtle is seriously endangered. Yet there is evidence that nearly one hundred species of plants unique to Peninsular Malaysia have become extinct during the past century as a direct result of forest loss, mainly in the lowlands of the west coast. Plans for any new parks or reserves in Malaysia should be based not only on the needs of large mammals but also on the distribution of plants, particularly those with a localized distribution and those needing special habitats. Apart from the emotional arguments for preventing extinctions, there are good economic ones, too, for conserving plants. For instance the value placed on those tropical hardwoods which come from natural forests is likely to remain high in the long term. Furthermore, tropical plants possess a store of currently un-identified natural products which may well become immensely useful, for example, as medicines.

Conservation and Development

How may the future of wild Malaysia be secured without inhibiting development? There are two things that we must be clear about. The first is that conservation and development are not incompatible. Perhaps that statement is too cautious. Any kind of development must take into account conservation principles. This is the concept behind a document called *The World Conservation Strategy*, produced a decade ago by a group of inter-national agencies concerned with environmental and nature conservation. The concept of development based on conser-vation principles has been stated many times before by many different people, but it must constantly be re-emphasized. The second point that must be made about conservation is that it involves everyone and every aspect of our environment. Concern for the survival of large animals is just one aspect. Loss of forest must occur, to produce food for people, commodities for export and land for towns and industries. But loss of forest has many effects: bad news for elephants, and bad news for people and development budgets if water supplies become muddy or erratic, timber supplies run down, rural people have no more forest produce and no money to buy substitutes from the towns. Slowly but surely, the full meaning of conservation is becoming realized and accepted. In 1988, an international organization known since its foundation in 1961 as the World Wildlife Fund officially changed its name to the World Wide Fund for Nature, in an effort to emphasize its true role. Other international agencies, including some of the development banks, are now taking serious notice of conservation issues. At the present time, the ninety-seven-nation Convention on International Trade in Endangered Species of Fauna and Flora (CITES) is banning the trade in ivory. It is also considering placing a prohibition on trade in tropical slipper orchids including many species in the genera *Paphiopedilum* and *Phragmipedium*, some of which are unique to Malaysia.

National Parks, State Parks and Other Reserves

Over the past three decades, genuine attempts have been made throughout Malaysia to strike a balance between the three main approaches to utilization of land and sea – between protection from exploitation, sustained harvesting of natural products and total replacement of the natural environment with some other form of land usage. Strict conservation is needed in order to protect those plant and animal species which cannot tolerate disturbance, to safeguard water supplies, to form a source of wildlife for recolonization into adjacent reserves, and to provide centres for research and recreation. This is achieved in Malaysia by the setting aside of national parks, state parks and other forms of protected reserve. Such areas occupy about five percent of the land area in both Peninsular Malaysia and Sabah, and two percent in Sarawak, but several further areas have been proposed, especially in Sarawak. The second form of land use is the basis of the timber industry in Malaysia, a key industry for both Sabah and Sarawak. Other examples of sustained harvesting include charcoal from mangrove forests, rattan from the forests, and fish from the sea and rivers. The third form of land use, replacement of the natural vegetation, in Malaysia mainly takes the form of oil palm, cocoa and rubber plantations.

Peninsular Malaysia has several excellent areas set aside for the conservation of wildlife, under the control of the Department of Wildlife and National Parks, a federal agency with offices in each state. Of particular importance is Taman Negara, Malaysia's

The Field Studies Centre at Danum Valley conservation area in eastern Sabah.

first and largest national park, which straddles the three states of Pahang, Kelantan and Terengganu. A very large percentage of Peninsular Malaysia's native wild plant and animal species occur within the borders of Taman Negara. The great attraction of this park is that it has never been disturbed by logging or cultivation, and there are no roads inside the boundary. It also contains the Peninsula's highest mountain peak, Gunung Tahan. Much smaller than Taman Negara, but nevertheless one of the finest protected areas in South-east Asia, is the Kerau Wildlife Reserve in Pahang. This reserve is particularly important because it contains a substantial amount of lowland dipterocarp forest in a region of moderately low rainfall. Everywhere else in Malaysia, indeed throughout South-east Asia, such forests have been cleared to make way for agriculture. Kerau also contains another high mountain, Gunung Benom. Further south, straddling the border regions of Pahang and Johor is a fine forest area which has been proposed as Endau-Rompin Park. Naturalists are especially concerned about protection of this area because it is the only remaining large tract of natural forest at the southern end of Peninsular Malaysia. A major scientific expedition organized by the Malayan Nature Society found several previously unknown and very rare species of plants and animals in Endau-Rompin, and the area also contains a breeding population of the endangered Sumatran Rhinoceros.

Peninsular Malaysia also has substantial areas of forest habitat protected in the form of forest reserves. Selective logging for timber is permitted in most forest reserves, but no form of settlement or agriculture is allowed. Thus the forest and its fauna remain, albeit altered by the effects of timber exploitation.

Malaysia has been blessed with an outstanding array of offshore islands. Most of the islands are characterized by fine sandy beaches, a little mangrove, coral reefs and a fairly sparse human population. Recognizing their special importance to fisheries, marine conservation, environmental quality and tourism, the Malaysian government has plans to protect the coral reefs and waters surrounding many of the islands as legally-constituted marine parks. At the present time, proposals for twenty-one such parks are under consideration.

In Sarawak, conservation and management of forests and of wildlife (except turtles, fish and other marine life) are all the responsibility of the Forest Department. Sarawak already has a diverse array of national parks and wildlife reserves, and proposals for many more are under consideration. Parks and wildlife reserves may possess a dual status of forest reserve. Most of the existing parks and sanctuaries were established some years ago, before biologists had come to realize that conservation of entire ecological systems necessitates large areas of natural habitat to be protected. Of the existing protected areas, only Gunung Mulu National Park and Lanjak-Entimau Wildlife Sanctuary are large enough to support permanent breeding populations of large animals. Like Peninsular Malaysia, Sarawak also has extensive forest reserves where exploitation of timber trees is permitted, but not agricultural settlement.

Sabah has one national park and five state parks, all run by a state government agency called Sabah Parks. Three of the parks consist mainly of hills and mountains, including Kinabalu Park which has the highest mountain and the greatest concentration and diversity of unique plants in South-east Asia. The other three parks are offshore islands and surrounding waters. Responsibility for wildlife everywhere outside these six parks lies with the state Wildlife Department. Well over forty percent of Sabah's forest land is forest reserve, under the management of the state Forest Department, including two wildlife reserves, where timber exploitation has been permitted but not agri-

culture. An organization called the Sabah Foundation, which is aimed at improving educational and other opportunities for Sabahans, has a vast, one-hundred-year logging concession in the hilly interior of Sabah. The foundation has established two large conservation areas (Danum Valley and the Maliau Basin) within its concession, for protection of flora and fauna, and for education and research.

FOCUS ON
PENINSULAR MALAYSIA

—◆—

Peninsular Malaysia consists of eleven states and the federal territory of Kuala Lumpur, in total occupying 131,235 square kilometres of land (50,670 square miles) with a coastline of 1,930 kilometres (1,200 miles). The commonly-used names of the eleven states, together with their correct, full names in parentheses, are as follows: Perlis (Perlis Indera Kayangan), Kedah (Kedah Darul Aman), Perak (Perak Darul Ridzuan), Penang (Pulau Pinang), Selangor (Selangor Darul Ehsan), Negeri Sembilan (Negeri Sembilan Darul Khusus), Melaka, Johor (Johor Darul Takzim), Pahang (Pahang Darul Makmur), Terengganu (Terengganu Darul Iman) and Kelantan (Kelantan Darul Naim). The 1988 census recorded 13.96 million people in the Peninsula.

The mountains and hill ranges, occupying about half of the land area, remain largely under forest cover, while most of the lowlands are cultivated with rice fields and plantations of rubber, oil palm and fruit trees. Peninsular Malaysia is one of the world's major sources of rubber and oil palm products, and downstream processing industries have been developed to produce items as diverse as surgical gloves, tyres, soaps and margarine. It is ironic to recall that in the Amazonian rainforest the rubber tree was once entirely overlooked, while the oil palm was merely a source of domestic oils from the rainforest for West African villagers. Today, both Malaysia's economy and millions of people world-wide depend upon the foresight of a handful of individuals in years gone by regarding the value of certain rainforest plants. Peninsular Malaysia also relies on mineral resources to sustain and diversify its economy. Tin, gold and other metals have been important in the past, but are less so nowadays, while petroleum and natural gas, obtained off the east coast, form the major source of Malaysia's wealth.

The north-western lowland region of Peninsular Malaysia, encompassing most of the states of Perlis, Kedah, Perak and the offshore island of Penang has experienced a long history of settlement and today is the most intensively cultivated in Malaysia. It is the only part of the country where rice is grown on a large scale, with two crops each year. The rest of the country relies on imported rice to sustain the urban population. A few natural areas remain in this region, mostly isolated hills. Kedah is perhaps the most ancient state in Malaysia, with the artefacts of Stone Age people, the remains of Hindu and Buddhist temples (see the section entitled *The Peoples of Malaysia*) and a present-day royal family which can trace its line back to those times. Perlis, the smallest state, situated in the north-western corner of Peninsular Malaysia, was originally part of the state of Kedah but was separated by Thailand in 1842 and later handed to British rulers. The beautiful Langkawi Islands lie off this part of the country. The state of Penang consists of an island together with a strip of land on the mainland opposite. Further south, Perak was settled from early times because of rich tin deposits in the fertile alluvial soils of the Kinta Valley. Little of the original wildlife remains in these unforested parts of Malaysia but there are various birds and mammals which adapt to habitats created by human activities. On dry land, mongooses may sometimes be seen, while on the coast and the lower parts of large rivers are the beautiful Smooth Otters (*Lutra perspicillata*).

Several essentially undisturbed forested hills do remain in this region, surrounded by the cultivated lowlands. Most famous is Gunung Jerai in Kedah, also known as Kedah Peak, which bears a microwave station on the top at 1,216 metres (3,990 feet) above sea level. On a clear day, the Langkawi Islands and Penang can be seen from this spot. The station can be reached by road and there is a rest house at 1,005 metres (3,300 feet). The slopes of Jerai are rich in native plant life, including orchids, rhododendrons and pitcher plants. The tree flora is typical of that found in Malaysian mountain forests (see the section entitled *Plant Life*), with representatives of the genera *Agathis, Dacrydium, Baeckea, Lepto-spermum, Podocarpus* and *Styphelia*. Visitors to the Peak are well-served by a system of hiking trails and picnic spots. South of the Kuala Kangsar – Matang road is Gunung Bubu, one of the most extensive forest reserves in the region. Almost all forest on flat land in this region has disappeared and with it, sadly, at least several tens of plant species. In Kedah, however, one of the largest palm trees in the world, called *gebang* (scientific name *Corypha elata*) may still be seen growing on bunds between the rice fields, even after hundreds of years of rice cultivation. In former times, both sago (edible starch) and paper were made from its massive stems.

In the past, charcoal made from trees cut from the coastal mangrove forests was the major source of domestic cooking fuel in Malaysia. Mangrove forest has been – and often still is – treated as wasteland, to be cleared and 'reclaimed' for 'development'. Fortunately, several decades ago, foresighted forest managers in Malaysia realized that mangrove forest, if harvested in a steady, sustained manner, could provide wood indefinitely, and some mangrove areas have been productively managed for many years. At Matang Forest Reserve, to the west of Taiping, mangrove wood to the value of up to US$9 million is harvested annually, while local fisheries which depend on the existence of mangrove bring in thirty million US dollars. It is estimated that 12,500 people have jobs ultimately dependent on the sustained use of this forest reserve. On top of this, Matang is used by about 50,000 migratory birds every year and is the only remaining site in Malaysia with a wild population of Milky Storks (*Mycteria cinerea*). Various other rare birds occur here, including Lesser Adjutant Storks (*Leptoptilos javanicus*), Chinese Egrets (*Egretta eulophotes*) and the Masked Finfoot (*Heliopais personata*). It is now known that the vast numbers of birds migrate annually back and forth between northern Asia and the tropical regions, to find

OPPOSITE PAGE A Silvered Leaf Monkey (*Presbytis cristata*) mother with her baby. Leaf monkeys, of which there are three species in Peninsular Malaysia, may be encountered in forest throughout the country. Feeding mainly on young leaves, seeds and flowers, they are most active in the early morning and late afternoon.

A view of the central part of Kuala Lumpur, at Jalan Sultan Ismail. In the 1850s this was a small settlement at the confluence of the Kelang and Gombak Rivers, in the heart of a tin-mining belt. The present-day city, a fine blend of old and modern, is the capital of the thirteen-state federation of Malaysia.

enough food for sustenance during the northern winter. Their migration involves a journey in excess of 5,000 kilometres (3,100 miles) each way. Many of the birds pass through Peninsular Malaysia, and some remain for several months, returning as the northern spring approaches. The west coast is particularly important for shorebirds such as sandpipers and plovers. Surprisingly, the mangrove islands off the mouth of the Kelang River – the muddy river which passes through the middle of Kuala Lumpur – together constitute one of the most important of all sites in Malaysia for these migratory birds.

A century ago, the region encompassing Kuala Lumpur and the states of Selangor, Negeri Sembilan, Melaka and Johor was one of scattered Malay communities along valleys in low, forested hill ranges and of extensive peat swamps. Now, there exists a diverse and productive array of towns, plantations and factories. Remnants of the original vegetation still exist, however, and some are accessible to the casual visitor. From Kuala Lumpur, a short drive will take us to the hill forest at Ampang (see page 97) or to the Batu Caves, situated in a limestone outcrop formerly threatened with quarrying, but now safe because of the site's cultural and biological importance. There are limestone

outcrops scattered through many parts of Malaysia, some of which are of great interest to naturalists because, in effect, they are 'islands', surrounded by a 'sea' of plantations or forests on substrates other than limestone. This means that they have rare or even unique plant species (see the Langkawi example on page 62). Many of these outcrops contain caves, and Batu Caves are a good example, albeit nearer to a metropolis than most of the others. Batu Caves represent the centrepiece of a fascinating story which clearly demonstrates the importance of an ecological way of thinking. In the section entitled *Plant Life*, we have described the importance of mangrove forests to fisheries and prawn catches, and also that controversial Malaysian fruit, the durian. Caves, mangroves, fisheries and durians are all linked by just one kind of animal: the Cave Nectar Bat (scientific name *Eonycteris spelaea*), which barely exceeds ten centimetres (4 inches) in length and which roosts in large colonies during the daylight hours in the dark parts of limestone caves. Every evening, around dusk, these bats stream out of the caves in search of food, which comes in the form of the nectar and pollen of flowers. Distance is of little object to the Cave Nectar Bat; it may fly several tens of kilometres each night. Different kinds of trees flower at different times of year, which is a good thing because this bat is thus ensured a supply of food throughout the year. The trees are also at an advantage because, in any one night, the bats feed on only one species of tree, transporting pollen between flowers and ensuring successful fertilization. At certain times of the year, the bats fly all the way to Kuala Selangor to pollinate

Tea estates in the Cameron Highlands.

the mangrove trees. The survival of mangrove forests, and of fisheries, is thus tied to the survival of these little bats. At other times of the year, the bats feed at durian flowers, and are responsible for the successful setting of durian fruits throughout much of the state of Selangor. This is a statistic not to be sneezed at: these durians provide great enjoyment to many people, including the growers, who collectively make over one million US dollars each year from their sales. When durian and mangrove trees are not in flower, the Cave Nectar Bats pollinate many other kinds of trees, including jackfruit, mangoes, jambu merah, petai and bananas, and a tree called *magas* (scientific name *Duabanga moluccana*) which grows on the sides of rural roads and river banks, thus reducing the amount of landslips and soil erosion.

Forests can be important in many ways which go unrecognized. The peat swamp forests of northern Selangor, often viewed as waste land after timber trees have been taken out, are now seen as a vital store of fresh water during dry periods, important for nearby communities and for rice fields. Streams in the swamps provide fish, and nowadays people drive from Kuala Lumpur, some eighty kilometres (50 miles) away, just to spend the weekend relaxing and fishing. Not far south of the peat swamp forest is an interesting development in the form of the Kuala Selangor Nature Park, which contains man-made lakes designed to attract waterbirds, with observation hides for visitors, adjacent to natural forest and mangrove. The area also provides habitat for Silvered Leaf Monkeys, otters and Adjutant Storks.

Heading south, to the states of Negeri Sembilan, Melaka and Johor, little natural vegetation can be seen from the roads. Pasoh Forest Reserve in Negeri Sembilan is not open to the casual visitor, but is of interest in that a great deal of basic biological research into how the Malaysian rainforests function has been carried out here over the past two to three decades. Mount Ledang (once known as Mount Ophir), an attractive small mountain near the Johor – Melaka border, retains much of its original vegetation and is readily accessible by road. Alfred Russel Wallace, originator along with Charles Darwin of the theory of evolution by natural selection, chose Melaka (then spelled Malacca) to start his eight-year sojourn in the eastern tropics. He visited Mount Ophir for a week in September 1854, his first experience of a tropical mountain, noting the occurrence not only of various plant species but also of tigers, rhinos and elephants – all three long since extinct from the area. The largest wild area in the southern part of Peninsular Malaysia – centered on Endau-Rompin (*see* page 105) – extends from Johor into Pahang. Off the east coast are many islands, of which Tioman is the largest.

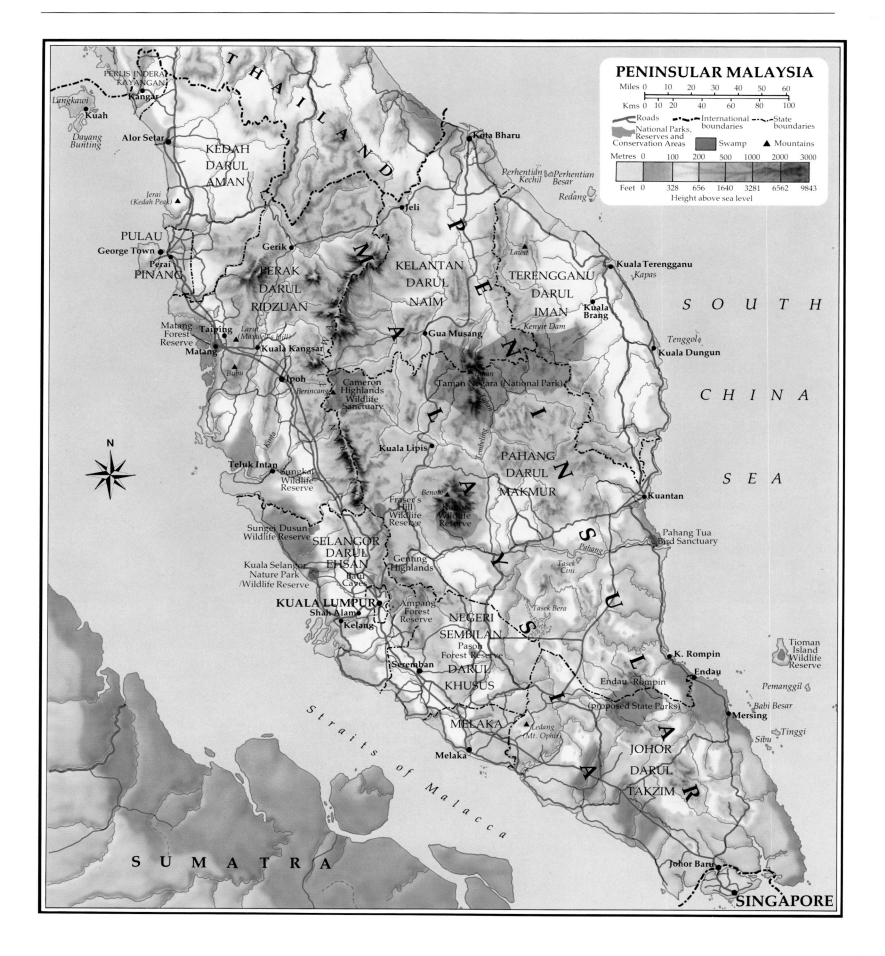

PENINSULAR MALAYSIA

Miles 0   10   20   30   40   50   60
Kms 0 10 20   40   60   80   100

Roads
National Parks,
Reserves and
Conservation Areas
International boundaries
State boundaries
Swamp
Mountains

Metres 0   100   200   500   1000   2000   3000
Feet 0   328   656   1640   3281   6562   9843
Height above sea level

THAILAND

PERLIS INDERA KAYANGAN
Kangar
Langkawi
Kuah
Dayang Bunting

Alor Setar

KEDAH DARUL AMAN

Jerai (Kedah Peak)

Jeli

PULAU
George Town
Perai
PINANG

Gerik

PERAK DARUL RIDZUAN

KELANTAN DARUL NAIM

Kota Bharu

Perhentian Kechil
Perhentian Besar
Redang

Lawit

Kuala Terengganu
Kapas

TERENGGANU DARUL IMAN

Kuala Brang

Matang Forest Reserve
Taiping
Matang
Larut (Maxwell's Hill)
Kuala Kangsar
Bubu

Gua Musang

Kenyir Dam

Tenggol
Kuala Dungun

Ipoh
Berincang
Kinta

Cameron Highlands Wildlife Sanctuary

Taman Negara (National Park)

SOUTH
CHINA
SEA

Teluk Intan

Kuala Lipis

PAHANG DARUL MAKMUR

Kuantan

Sungkai Wildlife Reserve

Benom
Fraser's Hill Wildlife Reserve

Krau Wildlife Reserve

Pahang
Pahang Tua Bird Sanctuary

Sungei Dusun Wildlife Reserve

Tasek Cini

SELANGOR DARUL EHSAN

Genting Highlands

Kuala Selangor Nature Park /Wildlife Reserve

Batu Caves

KUALA LUMPUR
Shah Alam
Kelang

Ampang Forest Reserve

Tasek Bera

NEGERI SEMBILAN

Pasoh Forest Reserve

Tioman Island Wildlife Reserve

Seremban

DARUL KHUSUS

K. Rompin
Endau

Endau-Rompin

Pemanggil

(proposed State Parks)

Babi Besar
Mersing

MELAKA

Ledang (Mt. Ophir)

Tinggi
Sibu

JOHOR DARUL TAKZIM

Melaka

Straits of Malacca

SUMATRA

Johor Baru

SINGAPORE

In the lowlands of southern Pahang lie Peninsular Malaysia's two main systems of freshwater lakes. The larger, Tasek Bera, is a swamp during dry periods while, at rainy times, it becomes an extensive array of large, shallow lakes. Tasek Bera is a traditional home of the Semelai Orang Asli, who have developed a way of life intimately bound up with the changing water levels. During wet times of the year, they are fishermen on the lake. When water levels rise very high, and the wild pigs of the region have been forced to congregate on patches of remaining high ground in the surrounding forest, the Semelai become harvesters of wild pork. What is of interest is that, according to Semelai custom, only large pigs may be killed, but not females with young, nor immature animals. If only modern societies would follow such examples . . . Also, according to Semelai tradition, pork obtained by one man or a group of men will be shared amongst all the families in the village. At drier times of the year, the Semelai shift from spending most of their time fishing and hunting to gathering produce from the surrounding forest. This is one of the regions where aromatic forest-tree resins are still collected to supply foreign markets. To the north of Tasek Bera, just off the great Pahang river, is a much smaller but permanent array of thirteen lakes called Tasek Cini (formerly Chini), which are more easily accessible to visitors.

The wildest part of Peninsular Malaysia lies in the centre and north-east, encompassing the inland parts of Kelantan, Terengganu and Perak, and Pahang to the north of the road between Genting Highlands and Kuantan. Most of this region consists of forested hill ranges and mountains, with agricultural plantations, orchards and rice fields in the lowlands and along the larger rivers. The region is bisected in a north-south direction by a railway and a sealed road. This is the most sparsely populated region in the Peninsula and the one where many traditional ways of life are retained. Peninsular Malaysia's two finest and most important established inland conservation areas, Taman Negara and the Kerau Wildlife Reserve, are situated here. Kerau contains the largest tract of totally protected lowland dipterocarp forest in Peninsular Malaysia. The western margin of this region is perhaps the grandest physical feature of Peninsular Malaysia: a range of mountains, which extends from Negeri Sembilan into Thailand, known as the Main Range or, in Malay, Banjaran Titiwangsa. The crest of the range peaks at between 1,400 and 2,000 metres (4,600 and 6,560 feet) elevation. Gunung Tahan, at 2,187 metres (7,175 feet) the highest mountain in the Peninsula, lies off the Main Range in the heart of Taman Negara. The quickest way to reach the mountain forests of this region from Kuala Lumpur is by driving up to the border between Selangor and Pahang, either to Fraser's Hill or Genting. Here, trees are smaller than those in the lowlands and foothills, and mosses and orchids are most abundant. Look out for the variety of attractive ground herbs, some of which can be found flowering at most times of the year. Birdwatchers will find a completely different array of species from those down in the lowland forests. As a bonus, walking along forest trails in the cool mountain air can be an exhilarating experience. Another accessible part of the region is the area known as Cameron Highlands, the temperate-climate, vegetable-growing area of Peninsular Malaysia, where extensive forest remains on the steeper slopes. Yet another way of seeing the grandeur of this region is by crossing from west to east along the highway between Gerik in upper Perak and Jeli in Kelantan.

Previously remote, many parts of this region are now being opened up with roads, some of which start as timber extraction routes and are subsequently maintained by the government to provide links between major settlements and plantations. The thriving town of Gua Musang in southern Kelantan, set in an area of scenic limestone outcrops, is fast becoming the hub of the region. Wild elephants, tigers and tapirs are still sometimes seen on roadsides not far from the town and there are hot mineral water springs in the forests. Week-long safaris led by Orang Asli and ending in a river journey on bamboo rafts are becoming popular with the adventurous. As the region becomes more accessible, new botanical discoveries and rediscoveries are being made. For example, a beautiful begonia (Begonia rajah, which has white flowers and bronze-green leaves flecked with darker tinged markings) was recently rediscovered – having been feared extinct for many decades – in the Terengganu forest at Gunung Lawit.

Several large dams have been constructed and are in the process of being planned for this central-northern region of Malaysia, both to generate hydro-electric power and to regulate floods. These dams lead to the flooding of quite large areas of natural habitat, and to the disruption of large mammal populations, but attractive forest-lined lakes have been formed in the process. Productive communities of freshwater fishes have been developed within the lakes, and the increasing popularity of sport fishing, both within Malaysia and elsewhere, has led to the initiation of an annual fishing competition at Kenyir Dam lake in Terengganu. Little-known but of interest to the naturalist in Terengganu and Kelantan are tracts of gelam (Melaleuca cajeputi) forest in swampy, freshwater land between the coastline and the inland hills – if carefully managed, these forests can supply an excellent durable hardwood and an aromatic, medicinal oil (known as cajeput) from the young leaves. It is also believed that the existence of the swamps acts to absorb flood waters and to prevent seepage inland of saltwater from the sea.

The coastline of north-eastern Peninsular Malaysia is blessed with long, sandy beaches and clusters of beautiful offshore islands. The beaches form the major turtle egg-laying areas of the Peninsula. The largest of all turtles, the Leatherback, weighing up to 500 kilograms (1,102 pounds) and measuring up to two metres (6½ feet) in length, comes up to certain sections of the Terengganu coast. Human activities – including excessive egg collection in the past and frequent deaths of adults in fishing nets – have made this creature one of the most endangered species in Malaysia (see the section entitled Wildlife Conservation). Fortunately, intensive joint efforts are now being undertaken by both government and non-government agencies to reverse the decline in numbers.

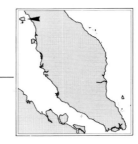

The Langkawi Islands

Situated off Perlis but part of the state of Kedah in the north-west corner of Malaysia are the ninety-nine Langkawi Islands. Largest of the group is Langkawi itself, dominated by the 911-metre (2,989-foot) tall peak of Gunung (Mount) Raya, near to the town of Kuah. Unlike the rest of Malaysia, the north-west normally has a pronounced dry season each year, and accordingly the flora is rather different from that elsewhere in the country. While most of the natural vegetation has been cleared on the mainland, a fair amount remains untouched on the Langkawi Islands. The oldest rocks in Malaysia, dating from the Cambrian period more than 500 million years ago, form outcrops on some of the islands. Limestone outcrops, geologically more recent, are another feature of the islands, which are particularly rich in flora associated with this rock. A beautiful little *palas* palm (*Liberbaileya gracilis*) grows only on limestone cliffs at the southern tip of Dayang Bunting Island, and nowhere else in the world. Zoologists have found at least five species of reptiles, eight of amphibians and two of butterflies which do not occur elsewhere in Peninsular Malaysia. It is said that in olden times, pirates made this area a base because they could evade detection in the complex scatter of islands. Through much of this century, the Langkawi Islands were mostly left out of the stream of developments occurring on mainland Malaysia. Well off the beaten track for travellers until a few years ago, the government now aims to make the islands a major tourist destination. This presents a challenging opportunity to combine development and conservation of wild species and natural scenery.

OPPOSITE PAGE The Langkawi Islands, viewed here from Pasir Hitam Beach, remained little disturbed by human activities, while just across the sea on the mainland a major centre of Malay civilization was developing.

BELOW About a thousand years ago, Indian and Arab traders would have passed close to the Langkawi Islands on their way to trade with what are now the states of Kedah and Perlis.

ABOVE The Langkawi Islands will see many changes as developers and tourists discover their beauty and come, inevitably, to alter the face of the islands for ever. One thing will remain unchanged: the tranquillity of sunset over the northern fringes of the Indian Ocean.

BELOW Tiny islets dot the landscape of the Langkawi group, forming a pristine refuge for plants, insects and birds. But, being so small, and continuously subject to a salty atmosphere, few forms of life can colonize them permanently.

ABOVE Unlike most of Malaysia, the Langkawi Islands are subject to seasonal fluctuations in rainfall. Botanists therefore find much of interest in the island forests.

RIGHT The freshwater 'Lake of the Pregnant Maiden' on the island of that name (in Malay, *Pulau Dayang Bunting*), the second largest of the Langkawi Islands. Legend has it that a Malay princess, forbidden to marry her lover, fled from the mainland to this lake and drowned herself. It is also said that barren women may become pregnant after bathing here.

LEFT A young Dusky Leaf Monkey (*Presbytis obscura*) in a papaya tree. This monkey is also known as the Spectacled Langur because, when mature, it has broad white rings around its eyes. Apart from the ubiquitous Long-tailed Macaque, this is the only species of monkey or ape on the Langkawi Islands.

BELOW LEFT A Soft-shelled Turtle. This species lives in the lower reaches of rivers. It lays its eggs in sandy banks.

BELOW CENTRE Lizards, like this *Liolepis belliana*, rely on warmth from the sun to keep their body temperature high enough to move about and obtain food in the form of insects. They are often especially abundant on islands, where the diversity of mammal and bird species is lower than on large land masses and where, accordingly, there is less competition for food.

BELOW RIGHT A caterpillar of a fruit-piercing moth of the genus *Othreis*, probably *Othreis homaena* (family Noctuidae). Moths of this group have brownish forewings and brilliant chrome-yellow hindwings. They are attracted to fruit, which they pierce with their proboscides to suck out juices. The striking caterpillars, with the curious hump at the rear and two large eyespot markings, feed on *Menispermaceae*. The chrysalids are formed between the folded leaves of the food plant.

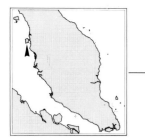

Penang

Previously a barely inhabited island, Penang was acquired in 1786 from the Sultan of Kedah by Francis Light, a private trader, on behalf of the British East India Company as a naval and trading base. The capital, George Town, was named after King George III, British monarch at the time. The island itself was initially named by the British after the Prince of Wales. Its later and present name refers to several species of palms, *pinang*, which grow both wild and cultivated in Malaysia. In 1800, the British acquired a part of the mainland opposite Penang, Province Wellesley, now Seberang Perai. But the British failed to protect Kedah, which was seized and ruled by Thailand from 1821 to 1909. Penang quickly became a regional metropolis in the nineteenth century and continued to develop steadily through the twentieth century. In 1985, Penang island was joined to mainland Malaysia by a 13.5-kilometre (8⅓-mile) long bridge. Today, the flat coastal land is heavily populated, while the hilly interior is covered in a mix of forest, orchards and vegetable gardens. Amongst the best of the remaining forest areas is the ten-square-kilometre (nearly 4-square-mile) Pantai Acheh Forest Reserve in the north-west corner of the island. The old and easily accessible Penang Botanical Gardens, with a well-established collection of local and exotic plants, both trees and shrubs, is a popular venue for both local and foreign visitors.

BELOW The sandy beaches of Penang island, (Pulau Pinang), are amongst the most heavily frequented stretches of coastline in one of the most populous parts of Malaysia. Yet the crowns of coconut and ru (or aru, *Casuarina equisetifolia*), seen here, are typical of coastal scenes throughout the country.

ABOVE Dawn is the coolest time to be around in Malaysia. On Penang island, especially, it is best to be up early in the morning to enjoy the natural beauty of the coastline.

OPPOSITE PAGE, LEFT Close interaction between monkeys is not confined to adult males and females. Adults may groom juveniles, as shown here, and they show great tolerance to youngsters who may insist on playing during siesta time.

OPPOSITE PAGE, RIGHT Like other animals, Long-tailed Macaques move around their habitat only when it is necessary and then usually to find food. Macaques feed on fruits, seeds, flowers, young leaves, small animals, insects and whatever humans care to provide for them. When replete, they rest and groom.

ABOVE Long-tailed Macaques (*Macaca fascicularis*), intelligent and adaptable, are most prominent amongst the wildlife species of Penang island, where this picture of an adult male and female pair was taken.

ABOVE Long-tailed Macaques are very social animals, which live in close-knit groups of up to thirty or so individuals. Here, a male grooms a female. This activity serves to build and cement bonds between individuals.

LEFT The strange purplish-black flowers of the Black Lily, *Tacca chantrieri*, a herb of shady parts of the rainforest.

MIDDLE LEFT The wild ginger, *Kaempferia pulchra*, is unusual amongst Malaysian herbs in shedding its leaves during dry spells. The underground parts of all *Kaempferia* species are commonly used in traditional Malaysian medicines.

MIDDLE RIGHT A wild forest ginger (probably *Zingiber spectabile*) stands out like a little sentinel in the gloom of the rainforest. The delicately-flavoured petals may be used as a flavouring, and an infusion of the leaves to cool inflamed eyes.

BOTTOM LEFT The same ginger flower viewed from above, showing its regular pattern of growth.

BOTTOM RIGHT The fruits of a *Knema* species – a relative of the nutmeg – exhibit one of the features shared by many rainforest trees: a dull-coloured, tough rind splits to reveal enticing scarlet flesh within.

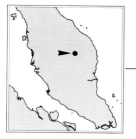

Taman Negara

Taman Negara ('National Park' in the Malay language) was the first area to be set aside specifically for conservation of wildlife and preservation of natural habitat in Malaysia. It incorporates land in the states of Pahang, Kelantan and Terengganu. Since its establishment in 1938–39, Taman Negara has remained the largest conservation area in Malaysia, one of the best-protected, and one of the finest in Asia. For this, Malaysians and the world owe much to a far-sighted man named Theodore Hubback, the first Chief Game Warden of the (then) Federated Malay States, who lobbied a reluctant colonial government for fifteen years, and all successive state and federal governments including the Department of Wildlife and National Parks. All have recognized Taman Negara as a key element in Malaysia's national heritage. Consisting entirely of land under natural forests and encom-passing 4,350 square kilometres (1,680 square miles) of land, mostly of rugged topography, Taman Negara ranges in elevation from less than 300 metres (1000 feet) to 2,187 metres (7,175 feet), at the top of Gunung Tahan, Peninsular Malaysia's highest mountain peak. The park represents the most important conservation area for a large proportion of Peninsular Malaysia's wild plant and animal species, including some which are found nowhere else, as well as elephants and tigers.

BELOW A view of the Tembeling River, one of the two main branches of the mighty Pahang River, near to Taman Negara. This scenery is typical of that on the way into Malaysia's National Park.

LEFT Much of the interior of Peninsular Malaysia looked like this many years ago. Most has now either been logged for timber or cleared for new settlement schemes. The landscape shown here is from Bukit Teresek, Taman Negara.

ABOVE The Tahan River in Taman Negara, with characteristic *neram* (*Dipterocarpus oblongifolius*) trees forming great arches over the water. This magnificent tree occurs only along river banks, mainly in Pahang and Kelantan, and is invariably draped with epiphytic plants like orchids and ferns – there may be more than a hundred species of epiphytes on just a few trees. *Neram* fruits are an important food for fish.

ABOVE, BELOW LEFT, BELOW RIGHT The Sumatran Rhinoceros
(*Dicerorhinus sumatrensis*) is one of Malaysia's endangered species,
reduced in numbers through hundreds of years of hunting to obtain
its horns. Taman Negara is one of its few refuges. Sumatran Rhinos
spend much of their time wallowing in mud pools, which they create
themselves by selecting a suitable spot and gouging out soil with their
horns. It is said that wallowing helps to keep off ticks and flies, but
probably its more important function is as a means of keeping cool
during the hot hours of the day.

ABOVE The Sun Bear (*Helarctos malayanus*) is the only species of bear in Malaysia, where it occurs in all kinds of forest habitats, including those in Taman Negara. Its alternative name is Honey Bear and, indeed, this bear does spend much of its time looking for the nests of wild bees.

RIGHT A White-handed Gibbon (*Hylobates lar*), one of three species of gibbon in Peninsular Malaysia, found in Taman Negara along with the larger, black Siamang. White-handed Gibbons occur in various colours, ranging from black to pale cream, but their hands are always white.

ABOVE Unlike their rhinoceros relatives, Malayan Tapirs (*Tapirus indicus*) do not wallow in muddy pools but instead keep cool by swimming in rivers and streams. But, like the Sumatran Rhino, the Tapir is a forest animal which feeds on leaves of small trees and saplings. The largest Malaysian population probably occurs in Taman Negara.

LEFT The Seladang (*Bos gaurus*) is the wild ox of Peninsular Malaysia. The bull (shown here) is a magnificent muscular creature, amongst Malaysian mammals exceeded in size only by the elephant. Of all members of the cattle family, Seladang are the best adapted to life in forests, feeding on saplings and shrubs as well as grass. They range over vast areas of Taman Negara.

LEFT The Sambar Deer (*Cervus unicolor*), known in Malaysia as *rusa*, is an adaptable species which can survive in most forest habitats, but prefers lightly wooded areas with gentle slopes and fertile soils. The bucks have antlers with a maximum of three points; a hind is shown here.

OPPOSITE PAGE, ABOVE A true forest-dweller, the Barking Deer (*Muntiacus muntjak*) is one of the few Malaysian mammals which seems to prefer steep hills to flat lands. This one is visiting a mineral lick in Taman Negara.

BELOW Various gliding animals occur in the Malaysian rainforests, like this lizard (*Aphaniotis fusca*) which takes advantage of the forest structure by gliding between the tree trunks a few metres above ground level. Then, standing on a trunk vertical to the ground, it looks out for and stalks insects.

BELOW The Black Giant Squirrel (*Ratufa bicolor*) is the largest tree squirrel in Malaysia, weighing one and a half kilograms (3¼ pounds) when mature. It spends most of its life in tall trees of the dipterocarp forest feeding on seeds and may be located by listening for its sharp, chattering call.

BELOW The Water Monitor Lizard (*Varanus salvator*) occurs in all lowland habitats where there is plentiful fresh water, feeding on carrion and any live animal prey it can find. This young individual was found on a tree buttress – it will grow to two metres (6½ feet) long when mature.

ABOVE A Grey-headed Fish-eagle (*Ichthyophaga ichthyaetus*) on a bank of the Tembeling River. This magnificent bird is a widespread resident of large watercourses in the Malaysian region.

BELOW The Garnet Pitta (*Pitta granatina*) is a beautiful ground-dwelling bird found in damp, shaded parts of the forest, where it feeds mainly on grubs, beetles and snails. Its call is a low, plaintive whistle.

BELOW The Yellow-crowned Barbet (*Megalima henricii*), one of several barbet species constantly heard but rarely seen. They are usually visible only when large numbers congregate to gorge on ripe figs.

ABOVE The Thick-billed Pigeon (*Treron curvirostra*) is another bird which congregates to feed on the fruits of large strangling fig plants, announcing its presence with flapping wing-beats and raucous, throaty gurgles.

BELOW LEFT The Green Broadbill (*Calyptomena viridis*) is a distinctive and beautiful bird of the lowland dipterocarp forests. Green feathers extend almost to obscure the broad beak which it uses to snap up fruit and insects.

BELOW The Oriental Cuckoo (*Cuculus saturatus*), a member of a large family of birds, all of which lay their eggs in the nests of other species. They are more easily distinguished by their call than their appearance.

ABOVE The mushroom shown here (a *Hygrocybe* species of the *firma* group) is short-lived, pushing up from the forest floor to disperse the millions of dust-like spores which spread the species. These spores are formed on the undersurface of the umbrella-like cap.

ABOVE Another *Hygrocybe*. A general term for the reproductive structures of fungi is fruit-bodies, although the spores are not fruits. Indeed they are far less complex than even the smallest seeds of flowering plants.

LEFT Some fungi, like this polypore (probably *Gloeoporus thelephoroides*), obtain their nutrients from dead wood. Fungi are thus important agents in causing the decay of fallen branches and tree trunks in the rainforest. Not only do they obtain nutrients for their own use, but they help to recycle them through the ecosystem.

RIGHT More *Hygrocybe* of the *firma* group. These and other fungi which arise from the ground obtain their nutrients from leaf litter and particles of plant material on or just below the forest floor.

ABOVE This *Podoscypha* (probably *P. involuta*) is another agent of wood decay in the forest. It differs from the polypore in the previous illustration in not having very small pores beneath the cap but simply wrinkles. The spores are formed on these irregularities, protected from the rain by the tough cap.

RIGHT Some fungal fruit-bodies, like this one (possibly *Microporus affinis*) are hard textured and last a long time. Beetles and other small creatures burrow into the flesh laying their eggs, the fungus offering protection to the hatching larvae which then feed on the tissues.

LEFT The Smaller Wood Nymph, *Ideopsis gaura* (family Nymphalidae). This beautiful butterfly has a curious, slow, flapping flight. Very similar in overall appearance to the related tree nymphs, or paper butterflies (*Idea*), these insects feed at the caterpillar stage on plants of the milkweed family.

BELOW The Great Egg-fly, *Hypolimnas bolina* (family Nymphalidae). This handsome butterfly is widespread in the Asian and Australian tropics. The male is always a deep, purplish blue-black on the upper side, with a large white area in the centre of each wing. Females, however, can vary considerably. The caterpillars feed on a variety of plants, including members of the *Acanthaceae*.

ABOVE The Red Helen, *Papilio helenus* (family Papilionidae). This fine swallowtail is widespread and often abundant throughout its range (from Sri Lanka to China and Timor). In Peninsular Malaysia it is more often encountered in the hills than in coastal areas. The caterpillars feed on plants of the orange family, including *Citrus*.

LEFT The Common Birdwing, *Troides helena* (family Papilionidae). These huge birdwing butterflies are amongst the most striking insects in the Asian tropics. After mating, the females seek out and lay their eggs on 'Dutchman's Pipe' vines (*Aristolochia*), the only plants on which the caterpillars will feed.

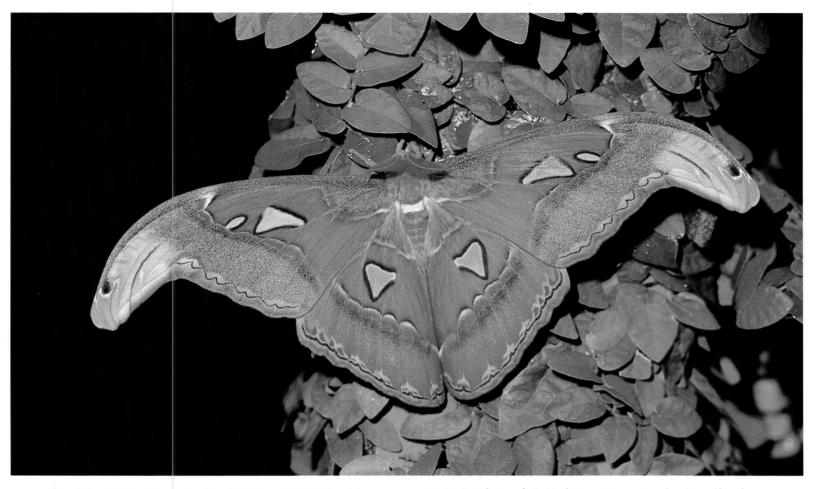

ABOVE The Atlas Moth, *Attacus atlas* (family Saturniidae). The Atlas, one of the largest moths in the world, occurs throughout India and South-east Asia. The striking spiny caterpillars feed on a variety of trees, including *Berberis* in some localities.

BELOW Rajah Brooke's Birdwing, *Trogonoptera brookiana* (family Papilionidae). This magnificent insect, with a wingspan of up to eighteen centimetres (7 inches), was first discovered by Alfred Russel Wallace and named in honour of Sir James Brooke.

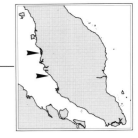

The West Coast

The natural coastline on the western side of Peninsular Malaysia is a mix of mangrove forests, sandy beaches and rocky shores. Of all the coastlines in Malaysia, this has been the most influenced and altered by human activities. Agriculture, plantations of various sorts, aquaculture, industry and housing have all made their mark. Various ways of using coastal land for human benefit have been pioneered in this region, starting long ago with irrigated rice fields behind the sand dunes, through use of mangrove forests for wood and fisheries, to rearing prawns and fish in brackish water ponds, and management of cockles in offshore mud. In recent years, it has even been found possible to grow oil palm – normally a dry-land crop – on salty mud irrigated with fresh water. Despite all this, the west coast still retains a good deal of interest in the way of natural scenery and wildlife. In the region as a whole, the productivity of marine life – fish,

prawns, crabs, molluscs, worms and so on – is high. Dependent on these for food are hundreds of thousands of water-birds of many species, some resident and some migratory, some rare and some abundant. Equally dependent, for jobs as well as food, are many tens of thousands of people who live in the region.

OPPOSITE PAGE The Silvered Leaf Monkey is the rarest species of monkey in Peninsular Malaysia, where it is confined to parts of the west coast. It may be seen most easily at Kuala Selangor, where the species has become tame as a result of legal protection.

BELOW Mangrove trees are the colonizers and rulers of coastal mud flats. Their aerial roots provide anchorage and a means of absorbing oxygen in the otherwise airless world below the mud.

LEFT Milky Storks (*Mycteria cinerea*), one of Malaysia's rarest bird species. The entire known Malaysian population of less than one hundred birds occurs at Kuala Gula in the Matang Forest Reserve.

OPPOSITE PAGE The White-throated Kingfisher (*Halcyon smyrnensis*), one of several attractively-coloured kingfisher species which may be seen on Peninsular Malaysia's west coast, and in coastal oil palm plantations.

BELOW LEFT The Crested Serpent Eagle (*Spilornis cheela*) is one of the most common and widespread birds of prey in Malaysia. It feeds on snakes and other small invertebrate animals.

BELOW The Lesser Adjutant Stork (*Leptoptilos javanicus*), a large and distinctive bird of the coastal regions of Malaysia. The largest known population occurs in Matang Forest Reserve.

LEFT A Painted Toad (*Kaloula pulchra*) near a west-coast village. The males of this small (75 mm/3 inch long) creature congregate and make loud honking sounds after heavy rain.

BELOW LEFT A Fiddler Crab (*Uca triangularis*) at Matang Forest Reserve. This is one of nearly one hundred species of fiddler crab which occur in mangrove habitats worldwide. Males have one enormously enlarged claw, used to intimidate and fight rival males, and to attract females for mating.

BELOW CENTRE A Fiddler Crab (*Uca rosea*) in the mangrove. The large claw of the male is also used to make a rapid, vibratory movement on the surface of the mud. This can be sensed by other crabs and serves, literally, to 'sound out' the mood of rival males lurking nearby.

BELOW RIGHT Mudskippers (family Periophthalmidae) are fish which spend much of their time out of water, on the mud of mangrove swamps. They move by pulling themselves forward with large pectoral fins. When alarmed, they skip away towards the nearest water or hole.

A dragonfly, *Crocothemis servilia* (family Libellulidae), occurring on the west coast of Peninsular Malaysia.

A dragonfly, *Orthetrum glaucum* (family Libellulidae) found in coastal areas and up to 1,400 metres (4,500 feet).

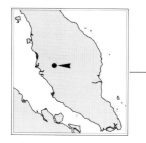

Cameron Highlands

The area known as Cameron Highlands, long accessible by road only from Perak and more recently from Kelantan, is actually in the north-western corner of the state of Pahang. The Highlands extend from about 600 metres (2,000 feet) above sea level to 2,031 metres (6,664 feet) at the peak of Gunung Berincang (= Brinchang). The area was discovered in 1885 by a government surveyor named William Cameron, who noted that the relatively flat area which forms the core of the Highlands (now known as Tanah Rata, or 'Flat Land' in English) would be ideal as a retreat from the hot lowlands and for growing crops of temperate climates. Enterprising Chinese settlers soon made their way there to grow vegetables and built a road to take their produce for sale in the lowlands. British planters moved in and built vacation houses. Later, tea plantations were established. Now, Cameron Highlands is the main area in Peninsular Malaysia for growing temperate-climate vegetables, flowers and tea. Produce from this area finds its way to all parts of Peninsular Malaysia and Singapore, and some of the tea is exported to other countries. Only fairly flat land is cultivated and the steeper slopes of the Highlands remain under natural montane forest, which can be explored from a number of hiking trails. The whole region was made a Wildlife Sanctuary in 1962.

BELOW Tree ferns (*Cyathea contaminans*) spring up in the mountains wherever there is a gap in the forest canopy.

LEFT The forest-clad mountains of the Main Range, seen from Gunung Berincang (= Brinchang).

BELOW One of the Malaysian balsams, *Impatiens oncidioides*, from Cameron Highlands.

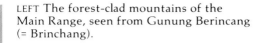

ABOVE *Oudemansiella canarii*, a widespread tropical fungus which is variable in size and shape but is always an active decayer of woody debris either when on the forest floor or even when still attached to trees.

BELOW The fruits of one of Malaysia's many attractive forest palms, *Pinanga polymorpha*, a species found throughout the hill ranges.

ABOVE A snake lily, *Arisaema filispadix*, from the Cameron Highlands. The rhizomes (fleshy stem bases) of these herbs accumulate starch to see the plant through difficult times, but animals are prevented from eating them by the presence of needle-like crystals.

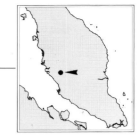

Fraser's Hill

Situated in Pahang next to the northern Selangor border, Fraser's Hill is named after an English adventurer, L. J. Fraser, who explored the area in the late nineteenth century. Fraser's Hill, which actually consists of seven hills at an altitude of between 1,220 and 1,525 metres (4,000 and 5,000 feet), is widely known within Malaysia as a recreational resort area. There is a variety of accommodation, some modelled on old English country houses and dating from colonial times, as well as a first-class modern hotel. Local and foreign visitors are encouraged to go to Fraser's Hill for its sporting facilities. But the area is also of great interest to naturalists, bird-lovers especially. The predominant natural habitat is lower montane forest (*see* the section entitled *Plant Life*), dominated by trees of the oak and laurel families. The bird fauna is especially rich because the area incorporates species from the lowlands, at the upper end of their altitudinal range, together with many of those species confined to true mountain forests. In addition, it has been found that Fraser's Hill is near the centre of a main 'flyway', or route, taken by birds which migrate annually to the Malaysian region in order to escape the winter of northern Asia. These birds are not the shorebirds which may be seen on Malaysia's coasts, but forest-lovers, including flycatchers, robins, thrushes, warblers and birds of prey.

BELOW A view of the Main Range from Fraser's Hill.

ABOVE A young Siamang (*Hylobates syndactylus*). This is the largest of the gibbon species. The powerful calls of the adult male and female pair are perhaps the single most characteristic sounds of the hill dipterocarp forests of Peninsular Malaysia; the species is absent from Borneo.

TOP RIGHT One of Malaysia's numerous species and forms of fig plant : *Ficus villosa*, on a rock face at Fraser's Hill.

MIDDLE RIGHT The fruits of an *Archidendron* tree, one of numerous little-known species of small tree which thrive in the Malaysian hill dipterocarp forests. The orange-red coloration of the pods probably attracts birds, but the means by which the seeds are dispersed is unknown.

RIGHT A ground-dwelling orchid, *Arundina graminifolia*, a species widespread in forests throughout Peninsular Malaysia. The form found on Fraser's Hill is considered to be amongst the most attractive.

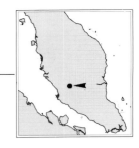

Genting Highlands

Genting Highlands, situated at the Pahang – Selangor border some forty kilometres (25 miles) to the south of Fraser's Hill is the nearest mountain resort to Kuala Lumpur. It is also by far the newest in Malaysia, having been opened in 1971. Like Fraser's Hill, it is best known for its recreational facilities, most especially as having the only gambling casino in Malaysia. The resort area, which has been pre-planned and is likely to expand still further, encompasses an altitudinal range of 1,036 metres (3,400 feet) (the golf course) to 1,800 metres (5,905 feet) at the highest peak. As at Fraser's Hill, nature lovers can find much of interest here. Not all recent developments at Genting Highlands have been sensitive to the needs of conservation. For example, the survival of two species of small tree of the citrus family (*Melicope suberosa* and *Maclurodendron magnificum*), known only from Genting High-lands, is at risk as a result of forest clearance. Fortunately, plants such as these do not need large areas in which to flourish, and the problem is one of lack of communication between naturalists and resort developers. Genting Highlands has been planned with imagination and deserves to succeed. But there is an important message here for 'conservationists' and 'developers' alike.

BELOW A striking view of the rainforest canopy from Genting Highlands. This forest is packed with all manner of plant species, some of which still remain unknown to science.

LEFT A view of the forest from Genting Highlands, showing the typical terrain of the Main Range.

BELOW A rare forest monitor lizard, *Varanus rudicollis* (shown here is a juvenile).

The Chocolate Tiger, *Parantica melaneus* (family Nymphalidae). This butterfly occurs at moderate altitudes in many parts of Peninsular Malaysia, where it can often be seen feeding at composites and other weedy flowers growing beside roads and tracks. In common with all milkweed butterflies, the males have a pair of eversible brushes, or 'pencils', at the tip of the abdomen, from which they can release scents essential for successful courtship.

A lacewing, *Cethosia penthesilea* (family Nymphalidae). Three species of lacewing are found in Peninsular Malaysia. Although attractive to look at, they emit a disagreeable odour if handled. The brightly-coloured red and brown caterpillars feed on species of passion flower plants and are thought to be chemically protected from some of their potential predators.

ABOVE LEFT This large mountain pitcher plant of Peninsular Malaysia (probably *Nepenthes sanguinea*) has brilliantly coloured lower pitchers. Nutrients from dead insects trapped in the pitchers are absorbed by the plant and help to supplement the poor nutrient supply of mountain forest soils.

ABOVE The flowers and developing fruits of a wild forest ginger, *Globba patens*. Like many other members of the ginger family in Malaysia, this is believed to have medicinal properties, and especially as a protective or tonic after childbirth.

LEFT A *Selaginella* species in Genting Highlands. Selaginellas are allied to the ferns, but are less widespread and usually confined to fairly damp, shaded spots.

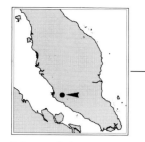

Ampang Forest Reserve

Most of the land in and around the Federal Territory of Kuala Lumpur was cleared of forest long ago. One of the few exceptions is Ampang Forest Reserve, retained to protect the catchment area of a dam ('ampang' means dam in the Malay language) which supplies water to a large part of the metropolis. The forest at Ampang, where a considerable diversity of wildlife remains, can be reached within an hour's travel from the heart of Kuala Lumpur. While much of the forest remains essentially undisturbed, parts were logged quite intensively in the past, and some interesting comparisons can be made. The differences in forest structure and composition of plant species are clear. Some two hundred bird species have been recorded within the reserve, a figure not far short of the maximum that can be expected within any patch of Malaysian rainforest. Some of these birds, however, are confined almost entirely to the tall, undisturbed

forest. The handsome, secretive trogons, for example, perch in shaded spots under tall trees, waiting for large insects to come by. Other birds, like some of the bulbuls, prefer secondary growth, where they actively seek small berries and insects. Only about fifty mammal species are known from Ampang, but more would undoubtedly be found with further study.

BELOW A Dusky Leaf Monkey (*Presbytis obscura*). Its local name, *chengkong*, is derived from the loud, nasal call of the adult male. Like other leaf monkeys, this species feeds on the leaves and seeds of trees and lianas.

RIGHT The Great Argus Pheasant (*Argusianus argus*), one of Malaysia's most spectacular but rarely-seen wild birds, found in hill dipterocarp forests throughout the country. The male, with tail feathers over a metre long, betrays its presence with loud 'wa-waauw' calls which carry far across the hills.

BELOW A young Malaysian Giant Toad (*Bufo asper*). When mature, this is the largest toad in Malaysia. The species is common along forest streams and may also be found in caves.

BELOW CENTRE A Copper-cheeked Frog (*Rana chalconota*), a species which frequents densely-forested stream banks in the hills.

BELOW RIGHT The Urania moth, *Micronia astheniata* (family Uraniidae). These moths occur from Sri Lanka to New Guinea. They are normally active at night, when they may be attracted to light. During the day they rest in low vegetation, from which they can readily be disturbed. The caterpillars feed under a silken web on *Eugenia* (Myrtaceae).

BELOW *Macaranga triloba*, a fast-growing tree which springs up at forest edges and clearings. There are many species of *Macaranga* in Malaysia, most of which are colonizers of gaps in the forest. Ants live inside the stems of some species, protecting the tree from leaf-eating insects.

ABOVE The seedling of a *Piper* plant on the forest floor. Most *Piper* plants have a distinct spicy flavour, and one member of the genus is pepper. All are graceful, slender climbers which require other plants for support.

BELOW *Selaginella willdenowii*, a fern-like herb with an unusual bluish tinge.

ABOVE A wild banana, *Musa gracilis*. The fruits of this and other wild bananas have large seeds and only a little, highly astringent flesh, eaten by birds and civets.

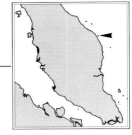

The Terengganu Coast

The long coastline of Terengganu offers magnificent stretches of long, sandy beaches, interspersed with patches of mangrove and nipa. This part of Malaysia, while always sparsely inhabited, has been a stronghold of Islam since well before the rise of Melaka in the early fifteenth century. Despite pressures over successive centuries from Melaka, Johor, Thailand and Britain, the coastal communities of Terengganu have retained an independent spirit. The harvesting of marine produce and coconut products have always provided many with their livelihood. Even now, villages and traditions seem much as they used to be, but there have been major changes since 1978, when large reserves of petroleum and gas were discovered offshore. From fields well over one hundred kilometres (62 miles) away from the coast, over 300,000 barrels of crude oil are extracted daily. Heavy industries have come in to make use of abundant cheap energy and more are likely to appear in the coming decade. Fortunately, the oil fields are well away from the beautiful offshore islands which dot the Terengganu waters. The giant Leatherback Turtles, which lay their eggs along only one small stretch of Terengganu coastline, are not so fortunate, and are now seriously endangered from a whole array of threats (*see* the section entitled *Wildlife Conservation*).

OPPOSITE PAGE A rocky part of the coast on Pulau Redang. Coral reefs with more than fifty-five genera of coral and over one hundred species of fish lie just off the island. Giant Clams (*Tridacna*), over-exploited elsewhere, are present and protected here.

BELOW An early morning view from Pulau Redang, one of Malaysia's beautiful offshore islands, fifty kilometres (31 miles) from Kuala Terengganu. There is one main fishing village on the island, which otherwise remains undisturbed.

ABOVE A sandy part of the coast on Pulau Redang. Hawksbill and Green Turtles lay their eggs here.

BELOW The clear waters of the South China Sea off the Terengganu coast.

ABOVE A Leatherback Turtle (*Dermochelys coriacea*) lays her eggs in the sandy beach at Rantau Abang in Terengganu. Each turtle comes up to lay between six and eight times during one season, usually between June and August, with an average of eighty-five eggs laid on each occasion.

RIGHT The Leatherback Turtle is now gravely endangered with extinction in Malaysia. All eggs laid are collected and allowed to hatch in a protected area on the beach at Rantau Abang.

BELOW The infant turtles are released on the beach, and they immediately head for the sea.

LEFT A male Rock Crab (possibly *Gecarcinus humilis*) on Pulau Redang.

ABOVE The 'Lantern Beetle' (family Fulfordidae), one of the strangest insects of the Malaysian rainforest.

LEFT The Malayan Giant Bull Frog (*Rana macrodon*). About ten centimetres (4 inches) long, this is Malaysia's largest frog.

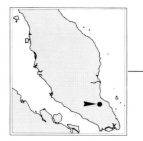

Endau-Rompin

In the southern part of Peninsular Malaysia there remains only one extensive forested area, a range of hills which rise to elevations of a little over 900 metres (3,000 feet). The northern parts drain into the Rompin River in Pahang, while the southern parts drain into the Endau River in Johor. It was realized during the mid-1970s that this area contained one of the largest surviving populations of the endangered Sumatran Rhinoceros (*see* the section entitled *Animal Life*) and, largely for that reason, Endau-Rompin, as the area became known, was proposed as a new national park. Due to complications regarding federal and state rights in forest land, this proposal did not materialize, but the states involved are seeking ways to ensure that the area is adequately conserved. In the meantime, the Malayan Nature Society organized a thorough exploration of Endau-Rompin during 1985–86, which involved more than seventy Malaysian scientists and nearly one thousand students and volunteers. Patron for the expedition was independent Malaysia's first prime minister, Tunku Abdul Rahman Putra Al-Haj. Sponsorship for this massive undertaking, the first of its kind organized, conducted and reported upon entirely by Malaysians, was provided by numerous local companies, organizations and individuals. During that expedition, the diversity of habitats and species within Endau-Rompin was found to be much greater than anticipated, showing that the area is of major conservation significance. For example, about twenty-five new species of plants were discovered, of which half appear to occur only within the proposed park.

BELOW Jasin River in the proposed Endau-Rompin Park.

OPPOSITE PAGE A waterfall on the Jasin River.

RIGHT The newly-discovered tree fan palm *Livistona endauensis*, which is believed to occur only in parts of the proposed Endau-Rompin Park.

BELOW The broad waters of the Endau River.

LEFT The pitchers of *Nepenthes* are sometimes very variable, those borne at the stem base being quite different from those of the aerial stems. This is an aerial pitcher of *Nepenthes rafflesiana* (compare with the illustration below).

MIDDLE LEFT In contrast to the upper pitcher, the lower pitcher of *Nepenthes rafflesiana* has broad wings; the variation in pitcher coloration within the same species can be remarkable.

BOTTOM LEFT The lower pitchers of *Nepenthes ampullaria*, a species of lowland swamp forest, are most distinctive; in favourable conditions the ground may be carpeted with pitchers.

TOP The fruit of *Freycinetia*, a woody climbing plant of the pandan family.

ABOVE The star-like flower of *Mussaenda mutabilis*, a woody climbing plant of the family Rubiaceae. It has a highly fragrant scent, in olden times used for scenting hair and clothes.

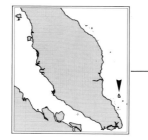

Tioman Island

Tioman Island is considered to be one of the most beautiful tropical islands in the world. Situated about forty-eight kilometres (30 miles) off the coast of Peninsular Malaysia and about four hours by boat from the coastal town of Mersing, Tioman is the largest of a scattering of sixty-four islands of volcanic origin in this part of the South China Sea. All the others are much smaller. At nearly thirty-eight kilometres (24 miles) long and nineteen kilometres (12 miles) wide, Tioman forms a microcosm of mainland Peninsular Malaysia several decades ago. The coast is fringed with clean sandy beaches and patches of mangrove, with coral reefs offshore. Although supporting both old settlements and a developing tourist industry, most of Tioman remains under natural forest, with clear streams and cascading waterfalls. Zoological expeditions are made to Tioman from time to time, and new species of invertebrate animals are still being found there. The forest is also a sanctuary for many kinds of plants and small animals found on the mainland as well, but under greater threat there from modern developments.

BELOW A forested stretch of Tioman Island's coastline.

OPPOSITE PAGE The coast of Tioman Island, showing coconut trees. There are several villages on the island, where many traditional ways of life are carried on.

LEFT Even when bathed in cloud, Tioman Island is easily recognized by its unique peaks, named Nenek Si Mukut and Batu Sirau.

BELOW LEFT A waterfall on the Ayer Besar River, Tioman Island.

BELOW An islet off Tioman Island.

ABOVE Many trees of the Malaysian forests carry their flowers directly on the trunk (cauliflory), the wild cempedak, *Artocarpus integer*, being a fine example. In this fruit tree, the male flowers are borne on sausage-shaped spikes.

ABOVE The torch ginger (*Etlingera elatior*), whose flower buds and unripe fruits are used as a flavouring in curries.

LEFT Most of Tioman Island is under forest, forming an excellent nature reserve.

BELOW *Lagerstroemia speciosa*, often planted as an ornamental tree, is native to the coastal lowlands of tropical Asia.

The female of a large thorny stick insect, *Heteropteryx dilatata*.

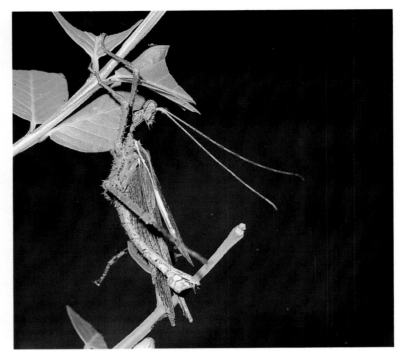

The male of a large thorny stick insect, *Heteropteryx dilatata*.

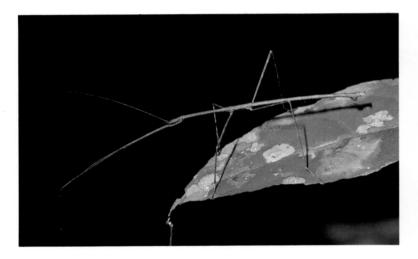

An unidentified male stick insect from Kerau Wildlife Reserve, Pahang.

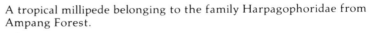

A tropical millipede belonging to the family Harpagophoridae from Ampang Forest.

Black Scorpions (*Palamneus sulpides*) occur in the lowland forests throughout Malaysia but are surprisingly inconspicuous.

A female leaf insect of the genus *Phyllium*, probably *Phyllium pulchrifolium*.

A Longhorn Beetle (family Cerambycidae), one of many thousands of species which occur worldwide.

A flower mantis (*Hymenopus coronatus*). Mimicking a flower, this insect remains disguised while it awaits other insects which become its prey.

Scavenging ants (*Camponotus gigas*) devour the carcass of a small rodent. Nothing is wasted in the rainforest.

A male Rhinoceros Beetle (sub-family Dynastinae), one of Malaysia's largest and most spectacular insects.

A spiny spider, *Gasteracantha arcuata* (family Araneidae).

A lichen spider of the genus *Pandercetea* (family Sparassidae).

The bizarre flowers of *Flikingeria fimbriata* last only for a day.

Dimorphorchis lowii has long inflorescences with two types of flower; this is an upper flower, while the basal flowers are completely different (*see* page 203, *below right*).

Luisia curtisii has cord-like leaves and small but pleasingly coloured flowers.

The orchid *Trichoglottis cf vandiflora*; the generic name refers to the hairy lip seen clearly in this photograph.

LEFT *Gastrochilus patinatus*, like many of its close relatives, has small but remarkably beautiful flowers.

Orchids have always tended to attract popular interest, because many have flowers of great beauty. The flowers of wild orchids vary greatly in size, according to species, but generally they are much smaller than those of domesticated hybrids and varieties. Thus, the beauty of wild orchids needs a close look to be appreciated. Many lovers of wild orchids have come to be fascinated not merely by the physical beauty of the flowers but by the way patterns, colours and shapes have evolved to attract the small insects which pollinate them. Most Malaysian orchids are epiphytes – they grow on tree trunks and branches. But there are also many species which grow on the ground. Of those, some

ABOVE *Liparis lacerata*, like most members of this large genus, has soft flesh-coloured flowers.

ABOVE RIGHT *Coelogyne asperata* is a robust plant, often occurring on steep embankments or cliff tops in the uplands.

RIGHT *Dendrobium linguella* occurs most frequently on riverside trees in the lowlands, where it flowers on bare stems.

have no leaves, and obtain their entire nourishment from decaying plant material underground. Despite their great variety, all orchids are defined by the basic structure of their flower. All have a total of three sepals and three petals, of which one – called the 'lip' – is of a shape totally unlike the others. In all orchid flowers, the male and female parts are on one 'column' which forms a part of the flower. In Peninsular Malaysia, wild orchids occur in forests from the coastal lowlands up into the mountains, with the greatest variety and abundance usually in the main hill ranges. Some hardy species colonize roadsides and the older trees planted in towns.

FOCUS ON
SARAWAK

◆

The modern state of Sarawak is based upon the river of that name which, together with Santubong, was in olden times the centre of a province of the Brunei sultanate. Sarawak now occupies 124,450 square kilometres (48,050 square miles) of land, and a large segment of the western side of Borneo, the third largest island in the world. Here are Malaysia's greatest remaining wilderness areas, with over seventy percent of the state remaining under forest cover of one sort or another. The landscape is dominated by flat plains and low hills near the coast, and rugged hills and mountain ranges in the interior. All of Sarawak's rivers drain into the South China Sea. The largest river system, and the largest in Malaysia, is the Rajang (also spelled Rejang). Although in terms of length the Rajang is dwarfed by the massive rivers of America, Africa and the Himalayas, it has been estimated that the volume of water flowing from its mouth is one of the greatest of any river in the world. This is because the rainfall is higher in Sarawak than in almost all other parts of the world, including the rest of Malaysia. Most of the country receives between 4,000 and 5,000 mm (150–200 inches) annually . It is the high rainfall which has contributed to leaching of minerals from the steep slopes in Sarawak and produced highly infertile soils. And, partly in consequence of this, the state's interior has remained lightly populated, and the rural peoples have been forced to rely on extensive shifting cultivation, whereby long periods of fallow between crops are essential to allow the soil to recover. Several interesting studies have been done of shifting cultivation in Sarawak. They indicate that, on average, about one hundred days of work are needed to cultivate one hectare of land, yielding about 800 kilograms (1,760 pounds) of rice, enough for one family for one year. Many shifting cultivators grow plants other than rice in their gardens. These plants include vegetables, tobacco, medicines and even natural poisons for use in obtaining freshwater fish.

Another major feature of Sarawak is the presence of extensive forested peat swamps on the coastal plains. Peat is formed by the accumulation of plant material – mainly twigs and leaves – under conditions of high natural acidity and where water does not drain freely. In Sarawak, the peat swamps have formed only during the past 5,000 years but are up to ten metres (over thirty feet) thick in some places. Strangely, valuable timber trees grow prolifically in some parts of this habitat, but the absence of true soil and extreme infertility mean that flowers and fruits are infrequent, while wildlife is very scarce. In the foothills between the main mountain ranges and the coastal swamps, and mainly in the southern parts of Sarawak, people have practised shifting cultivation to such an extent that much of the original forest has been replaced by scrub and grassland.

Pepper, a crop which can tolerate high rainfall, is grown on more fertile slopes, and Sarawak has for long been one of the world's major producers of black and white pepper. Other crops grown include rubber, coconut, oil palm, cocoa, coffee and tea, but all are on a small scale in comparison to other parts of Malaysia.

Timber and wood products from the natural forests are currently bigger income-earners for the state. In terms of the state's overall economy, however, petroleum and natural gas, all from offshore fields, are of prime importance. Sarawak also has the largest known reserves of coal, bauxite and kaolinitic clay in Malaysia. Gold, from the Bau area not far from Kuching, was one of the reasons that early Chinese settlers came to Sarawak River, but now very little remains.

All the rural people of Sarawak have a way of life strongly tied to the natural environment, and a culture rich in legends, beliefs and materials (wood, bark, animal products and so on) from the forest. The non-Moslem natives of Sarawak are often referred to as Dayaks, a term which has widespread acceptance but which obscures the existence of several distinct peoples. In terms of numbers, the Ibans of the southern and central parts of the state are by far the most numerous (thirty percent of the state's population), although it is believed that they are relatively recent arrivals to the island of Borneo, possibly originating only a few hundreds of years ago from Sumatra. The Bidayuhs (eight percent of the population) are concentrated in the southern hills and plains, while the Kenyahs, Kayans, Kelabits, Penans and others (about five percent in total) occupy the northern river valleys and hill ranges. The Moslem Malays and Melanaus (about twenty-six percent) live predominantly on the coast and lower flatlands. The balance of the population (about twenty-nine percent) is composed mainly of Chinese people, who are concentrated in the towns and larger villages. Rural communications in Sarawak are still the poorest in the country because of the rugged nature of the terrain and the sheer distances involved. The extent of the rural road system is limited, and most long-distance travel in the interior is done either by boat, in the regions with large rivers, or on foot or by air.

Sarawak has one of the most extensive systems of parks and sanctuaries of any of the states in Malaysia, although the sizes of some of these protected areas are considered rather small in the light of modern knowledge of tropical forest systems. Several of Sarawak's national parks are open and accessible to the general public and three of them in particular – Bako, Niah and Gunung Mulu – show much of what natural Sarawak is really like. Other existing national parks are Lambir Hills, Similajau and Gunung Gading. There are two wildlife sanctuaries important for primates – Samunsam (for Proboscis Monkeys and species of the coastal swamps) and Lanjak-Entimau (for Orang-utans and species of the hill forests). During a survey of Lanjak-Entimau involving the Royal Malaysian Air Force, the Sarawak Forest Department and WWF Malaysia, it was found that the presence of Orang-utans can be detected by spotting their nests from a

OPPOSITE PAGE The Rhinoceros Hornbill (*Buceros rhinoceros*). This magnificent bird forms the basis of the state emblem of Sarawak. Known in Sarawak as *kenyalang*, it plays a major part in a traditional Iban religious ceremony, *gawai kenyalang*. Pairs of this hornbill species occupy territories in the dipterocarp forests throughout Malaysia, where their loud 'gronk' calls are a characteristic sound.

Sunset over the Sarawak River at Kuching.

helicopter. This technique was used subsequently elsewhere in Sarawak and Sabah to plot the distribution of Orang-utans in remote areas, and later a method was developed to estimate the numbers of Orang-utans present from the number of nests seen. Immediately to the south of Lanjak-Entimau is Sarawak's only large man-made lake, in the headwaters of the Batang Ai, formed by a hydro-electric power dam. A number of additional parks and sanctuaries have been proposed.

At 2,423 metres (7,950 feet) the highest mountain in Sarawak is Murud, near the border with Kalimantan (Indonesia). It is in this mountain range that all the three large rivers of northern Sarawak – the Baram, Limbang and Trusan – have their sources. Also in this area is Batu Lawi, a rather slender, isolated sandstone peak which rises to over 2,000 metres (6,500 feet), way above the main forest canopy below. And in this region, too, small numbers of the Sumatran Rhinoceros were recently discovered to survive, nearly forty years after the species was feared extinct in Sarawak. South of Murud, in the very upper reaches of the Baram river, is the old mountain valley settlement of Bario (also spelled Bareo), inhabited by the Kelabit people. Until the advent of aeroplanes, this was one of the remotest settlements in Borneo. It is intriguing that the wet rice grown here appears to have been developed from hill rice independently of wet rice elsewhere. A study of the hunting of wildlife in the upper Baram River in 1984–85 found that Bearded Pigs formed the major

source of meat and fat for the rural people. These pigs may cover enormous distances within the space of a year in search of food. When the dipterocarp forests are full of fruits the pigs remain in the lower hills and valleys, but when the montane forests are producing acorns they move many tens of kilometres up into the mountains, sometimes to the Kalimantan border. The same study estimated that nearly 20,000 metric tons of wild animal meat are consumed annually in Sarawak. Nowadays, much is taken down-river to be sold in the towns. This study helps to demonstrate how the value of wildlife is often overlooked, both directly in terms of money value, and in terms of the health and sustenance of rural communities. The lower Baram region contains several natural freshwater lakes, of which the largest is Loagan Bunut on the Tinjar tributary.

South of the Baram River, separating this mighty river system from the even larger Rajang, is a rugged, broken plateau called Usun Apau. Much of this area consists of old volcanic rock, which typically produces fertile soils, but the soils here are highly infertile because of high silica content and heavy rainfall. Peat has accumulated on many parts of the plateau. In consequence, the vegetation on Usun Apau resembles heath forest. The high, rugged area between Usun Apau and the Kalimantan border is perhaps the most remote and least-explored in all Malaysia and, for that matter, in Borneo too.

The visitor who wishes to explore rural Sarawak without mounting a major expedition has various options, but all expeditions should include a river journey. The trip to Gunung Mulu National Park includes a journey on the lower Baram and Tutoh

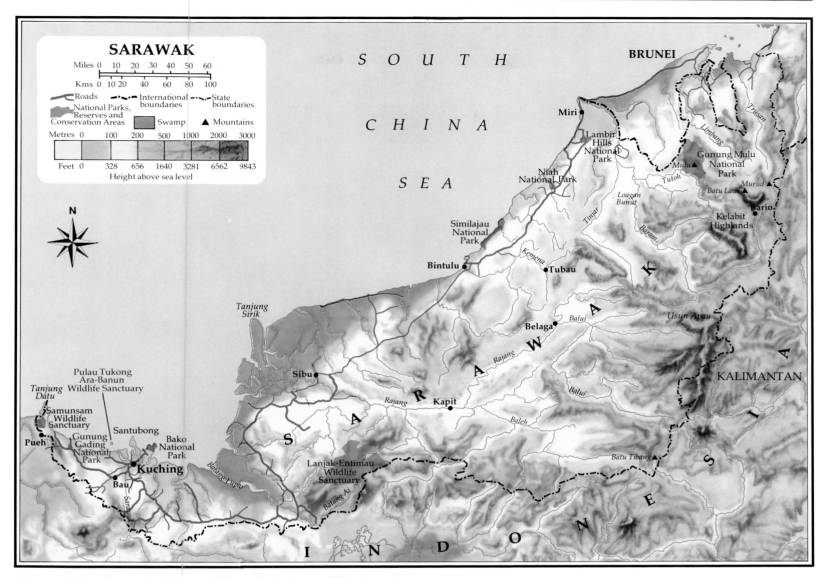

SARAWAK

Miles 0 10 20 30 40 50 60

Kms 0 10 20 40 60 80 100

~~~ Roads -·-·- International boundaries -··-··- State boundaries

National Parks, Reserves and Conservation Areas ▨ Swamp ▲ Mountains

Metres 0 100 200 500 1000 2000 3000

Feet 0 328 656 1640 3281 6562 9843

Height above sea level

N

SOUTH CHINA SEA

BRUNEI

Miri •

Lambir Hills National Park

Niah National Park

Trusan

Limbang

Mulu ▲ Gunung Mulu National Park

Tutoh

Batu Lawi ▲ Murud ▲

Bario •

Kelabit Highlands

Loagan Bunut

Tinjar

Baram

Similajau National Park

Kemena

Bintulu • • Tubau

Tanjung Sirik

Usun Apau

Belaga • Balui

KALIMANTAN

Rajang

Sibu • Kapit •

Balui

Rajang

Baleh

Pulau Tukong Ara-Banun Wildlife Sanctuary

Tanjung Datu

Samunsam Wildlife Sanctuary

Santubong

Pueh • Gunung Gading National Park Bako National Park

Kuching

Bau •

Batang Lupar

Lanjak-Entimau Wildlife Sanctuary

Batang Ai

Batu Tibang ▲

Sarawak

INDONESIA

Rivers. A simpler option is one of the relatively small but broad rivers which drain the south-western end of the state. One of these, the Batang Lupar is still famed for its occasional man-eating crocodiles. Local belief is that one or two enormous crocodiles, half real, half myth, are responsible for the disappearance of people. But studies have revealed that crocodile attacks are seasonal, during the period when mating is believed to occur, and most likely all mature males are aggressive and potentially dangerous at that time. An alternative river trip could cover sections of the lower or middle stretches of the Rajang. A more ambitious trip, which avoids having to cover the same ground twice, is to take a boat from Bintulu to Tubau, in the upper part of the Kemena river. From Tubau, an overland trek leads to Belaga, on the Rajang river. This is an area inhabited by the Kayan people. From Belaga, a boat may be taken up the Balui river, which flows from the Usun Apau plateau, through the Bakun rapids. Back at Belaga, another boat can be taken all the way down the Rajang river to Sibu, passing through the Pelagus Rapids and Kapit, an area of the Iban people. Occasionally, groups take off for the Baleh or Balui tributaries, in the upper reaches of the Rajang drainage, which have their source on Batu Tibang near the Kalimantan border. Such expeditions, though offering much interest and excitement, are expensive to mount and take an unpredictable amount of time.

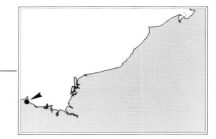

Samunsam Wildlife Sanctuary

Samunsam Wildlife Sanctuary is a hidden gem in Sarawak's system of conservation areas, only open to visitors on special request, less widely known than the national parks and lacking their spectacular physical scenery. Instead, Samunsam is complementary to the other parks and sanctuaries, containing a rather different array of habitats and offering a different atmosphere. Situated near the westernmost corner of Sarawak, the boundary of the sanctuary extends from behind the sandy coastline up to the border with the nearby province of West Kalimantan in Indonesia. The Samunsam River runs through the middle of the sanctuary, but its headwaters are outside in the Pueh mountain range. The lower reaches are fringed with mangrove and nipa. Much of Samunsam's 6,090 hectares (15,048 acres) consists of dry lowland forest on flat and moderately sloping land, and away from the coast and main river a kind of tall heath forest dominates. Unlucky is the observant visitor who fails to see a few wildlife species during a short visit. Bearded Pigs, Barking Deer, monkeys, squirrels and treeshrews are likely to be seen during the day, while birdwatchers have a good chance of seeing the rare Wrinkled Hornbill. Samunsam is the location of the only detailed study of Proboscis Monkeys to have been carried out in Malaysia.

BELOW A mangrove tree (*Rhizophora mucronata*), showing an unusual feature in the plant kingdom: the long pod-like appendages are actually roots, which develop from the small fruits of this species before they detach themselves from the tree. The root may reach a length of forty-five centimetres (18 inches) on the parent tree, with tiny leaf buds at the top end. Some of these 'seedlings' develop beneath the parent tree, but many are dispersed by sea currents.

ABOVE The eggs of a Black-naped Tern (*Sterna sumatrana*) on a rocky islet off Samunsam. This sea bird breeds on many such sites off the west coast of Borneo.

ABOVE RIGHT A mudskipper at Samunsam. In any one place, several species of mudskipper fish co-exist, some feeding on micro-organisms, algae and small plants, some on crabs, and others on a mixed diet of plants and animals.

RIGHT A crab (*Grapsus strigosus*) on a rocky islet off Samunsam.

RIGHT A Hermit Crab, one of several species of crab which do not have a natural hard carapace to protect the rear part of their bodies. Instead, they 'wear' empty mollusc shells.

BELOW Riverine forest at Samunsam, typical of the lower reaches of rivers throughout Borneo, showing three distinct palm species: nipa (*Nypa fruticans*), nibong (*Oncosperma tigillarium*) and rattan (probably *Daemonorops longispatha*).

OPPOSITE PAGE The transitional forest zone between swamp and seasonally-flooded dry land. The exposed roots allow the trees to obtain extra oxygen, in short supply beneath the waterlogged soil.

OPPOSITE PAGE Proboscis Monkeys (*Nasalis larvatus*) in riverine forest. This species is unique to Borneo, where it is found only near rivers and coasts.

RIGHT An immature male Proboscis Monkey, showing the strange nose which will become much larger and pinker as the animal grows older.

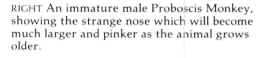

BELOW A male Proboscis Monkey. These monkeys feed primarily on leaves and seeds in forests never more than a kilometre from the sea, or from a large river or lake.

BELOW RIGHT Proboscis Monkeys are bulky, but they spend most of their time in the trees. They can make prodigious leaps between trees, but are also adept at swimming across rivers where necessary.

LEFT A fan palm (*Licuala* species) in the lowland forest at Samunsam.

BELOW A *Cymatoderma* species photographed in Samunsam Wildlife Sanctuary where it, and related tough and leathery fungal species, play an important role in breaking down the tissues of the forest giants.

RIGHT A cycad (*Cycas rumphii*) – a living plant fossil, the ancestors of which evolved long before any flowering plant, and which graced the landscape during the time of the dinosaurs. This cycad reaches only a few metres high, but some specimens are over a thousand years old. Their natural habitat now is sandy coasts.

BELOW The flower of a nipa palm being pollinated by insects.

BELOW RIGHT A flowering *Poikilospermum* (family Urticaceae), a soft, woody climbing epiphytic plant of the lowland forests and old gardens. The stems and roots contain large quantities of drinkable water.

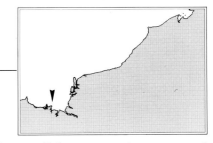

Bako National Park

Bako National Park is the oldest strictly-protected conservation area in Sarawak, having been established in 1957, six years before the state's independence through Malaysia. It is also the smallest park in Sarawak and the only one to have been established by the colonial government. In the early 1960s, the world's first pioneering attempts to reintroduce captive Orang-utans back into the wild were made here by Barbara Harrisson, wife of the then curator of the Sarawak Museum. The experiment was moved to Sabah when it was realized that the limited extent of forest locally made success impossible. Situated on the coast, to the north-east of the capital town of Kuching, Bako is the most easily accessible national park in Sarawak. The juxtaposition of several natural circumstances – proximity to the sea's waves and salt, a very heavy annual rainfall, and highly infertile sandstone substrates – has endowed Bako with an extraordinary array of habitats and special plants. All these are within an area of 2,727 hectares (6,738 acres), and all are visible from a well-maintained, thirty-kilometre (19-mile) system of trails. Access to the park is by road, followed by a short boat journey, or alternatively a longer boat journey from Kuching.

OPPOSITE PAGE Tree roots on a forest path in Bako National Park. This picture illustrates how trees manage to extract minerals sufficient for their needs from even the infertile sandy soils of Sarawak, and how they check erosion under the conditions of heavy rainfall which prevail in this region.

BELOW A view at Bako National Park on the coast of southern Sarawak. This fine park contains a remarkable array of habitats and wild species within a small area well served by clear trails.

ABOVE A waterfall and freshwater pool in Bako National Park.

ABOVE A view towards Santubong from Bako National Park.

OPPOSITE PAGE Sandstone cliffs at Bako National Park.

BELOW *Dischidia rafflesiana*, a slender climber with white latex. It has both typical leaves and special, expanded, hollow leaves (shown here) about three centimetres long and yellowish-green in colour. Ants shelter inside, leaving their excreta as nutrient for the plant.

BELOW A fan palm (*Pholidocarpus maiadum*) in the forest at Bako.

ABOVE A young Leopard Cat (*Felis bengalensis*) photographed in the forest at dusk. This beautifully-marked small wild cat thrives in a variety of forest and cultivated habitats in Malaysia.

RIGHT A Long-tailed Macaque (*Macaca fascicularis*), one of three species of monkey which can be seen at Bako.

LEFT The Colugo (*Cynocephalus variegatus*), also known as the Flying Lemur, is one of the most unusual Malaysian mammals which occur at Bako. With teeth formed like tiny combs, and a diet which remains unknown, it has no close relatives in the animal kingdom. Its mottled coloration provides some camouflage.

ABOVE One of the Colugo's most prominent features is the presence of membranes between the limbs and tail, which enable it to glide between trees to a distance of about a hundred metres (330 feet). Flying squirrels, superficially similar, differ in having the tail completely free of any membrane.

LEFT A red-coloured land crab (probably *Perbrinckia loxopthalma*) at Bako.

ABOVE An attractively-marked skink (*Dasia vittata*), one of a variety of lizards which occur at Bako.

Santubong

The Santubong region – a peninsula to the north of Kuching, one hour away from the capital by fast boat – evokes thoughts and images of days gone by. Mount Santubong, at the end of the peninsula, reposes steady and tranquil on fine days, or looms moodily during storms or rain. This part of the Borneo coast was a landmark and stop-off point for early traders and explorers. Coins, pottery fragments and even a statue of Buddha have been found at Santubong, dating from the T'ang and Sung dynasties, which flourished during the period 620 to 1280 AD. There may then have been a major settlement here, either of Chinese immigrants, or of settlers from the Sumatra-based, Buddhist Sri Vijayan empire. In the fourteenth century, the Java-based Hindu Majapahit empire extended its influence to the west coast of Borneo, and carved rocks from this period have also been found at Santubong. It is likely that even earlier, in the third and fourth centuries, Chinese traders and Buddhist pilgrims stopped here for fresh water and food on their way to India. Much later, and up to the middle of the present century, European adventurers and administrators would have passed Santubong by boat before entering the Sarawak River on their way to Kuching. In 1839, a young Englishman named James Brooke left Singapore and moored his ship at Santubong. A few years later, after helping the rulers of the Sarawak River district to quell unrest and piracy, he was made Rajah of what is now the state of Sarawak.

OPPOSITE PAGE A view from Santubong, showing the Sarawak coastline and striking cumulo-nimbus clouds. This region has the highest rainfall of any coastal part of Malaysia – an annual total of 4,500 millimetres (177 inches).

BELOW Nipa palms and mangrove trees line the Sarawak River.

LEFT The slopes of Mount Santubong remain cloaked in forest, just as they appeared to travellers centuries ago.

BELOW The Blue Pansy, *Precis orithya* (family Nymphalidae). A lowland butterfly, the males are often encountered in open areas. The dark, spiny caterpillars of this widely distributed species can be found on *Hygrophila* (Acanthaceae).

BOTTOM OF PAGE A dragonfly, *Neurothemis terminata* (family Libellulidae).

Niah and Lambir Hills National Parks

Niah and Lambir Hills are two small parks centered on hills in the coastal lowland of northern Sarawak and accessible from the town of Miri. The importance of Niah lies in its large limestone outcrop and caves, which not only support a specialized cave fauna, but also represent the major archaeological site of Borneo. In contrast, Lambir Hills are primarily of botanical interest, with scenic forest and streams in a region of sandstone.

Niah National Park headquarters and hostel are accessible by road. The caves are reached via a three-kilometre raised board-walk through pleasant lowland rainforest rich in plant species and bird life. Human beings have lived in or near the caves for an estimated 40,000 years. Excavations inside the caves, done mainly during the 1950s, revealed a remarkable, continuous sequence of human and animal remains up to recent times.

People evidently hunted a wide array of animals for food including Orang-utans, which appear to have been the second most popular prey species after wild pigs. In former times, there were also giant pangolins (now totally extinct) at Niah, along with mammal species now confined to high mountains. The present cave fauna consists of swiftlets, which produce edible nests, and a variety of bats, plus many invertebrates which live on the waste products and dead bodies of these animals.

BELOW A view of the forest from Niah Caves.

RIGHT The roots of a large fig plant (*Ficus* species) at Niah National Park. The crown of the fig plant is shading seedlings on the forest floor and keeping them moist. Exposed to direct sunlight, they would die.

BELOW Niah Caves – the location of Borneo's most important known archaeological site. Remains of animals and of human activities dating back to about 40,000 years ago have been found and documented here.

ABOVE A Bay Owl (*Phodilus badius*), a handsome nocturnal bird of the lowlands of Malaysia. It feeds on small birds, rats and insects.

ABOVE The Barred Eagle-owl (*Bubo sumatranus*), Malaysia's largest strictly nocturnal bird. It feeds on rodents, small birds and reptiles, sometimes inside the caves at Niah.

BELOW The Crested Wood Partridge (*Rollulus roulroul*), a bird which is usually seen on the ground in the lowland forests of Malaysia. Living in small groups, they feed on fallen fruits and insects.

BELOW The Red-headed or Ashy Tailorbird (*Orthotomus ruficeps*), a common, small bird of the lowlands throughout Malaysia. Most abundant in secondary forests and forest edges, it feeds on insects.

ABOVE The Great Woolly Horseshoe Bat (*Rhinolophus luctus*), which roosts in caves and crevices, emerging after dark to catch insects.

LEFT The Bornean Tarsier (*Tarsius bancanus*), a strange nocturnal primate which feeds on insects and small vertebrate animals.

ABOVE The Common Green Agamid Lizard (*Calotes cristatellus*), a species of forest edge and rural gardens.

FAR LEFT A long-legged cave centipede, or scutigerid, one of the many invertebrate animals which spend their life inside caves.

LEFT A forest bug (order Homoptera) which is disguised as a small, white feather.

OPPOSITE PAGE A waterfall in Lambir Hills National Park.

RIGHT A view of the forest canopy in Lambir Hills.

RIGHT *Bauhinia*, a woody climbing plant of the legume family. It 'climbs' by putting out small, soft hooks, which grasp on to other plants for support, and then curl and develop woody tissue to hold on permanently. Bauhinias, of which there are several wild species in Malaysia, are often common at forest edges.

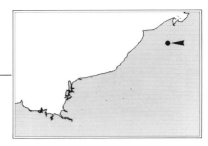

Gunung Mulu National Park

Pride of place amongst Sarawak's national parks must go to Gunung Mulu, which, within its 52,887 hectares (130,683 acres), has one of the most fascinating mixes of natural habitats and spectacular wild scenery in South-east Asia. The highest peak in the park, Gunung Mulu, at 2,377 metres (7,798 feet) above sea level the second highest in Sarawak after Murud, is part of a range of sandstone and shale hills. Other hills in the park, although smaller, represent some of the most massive limestone outcrops in South-east Asia: Gunung Api (1,750 metres or 5,740 feet), Gunung Benarat and a range of hills to the south (about 450 metres or 1,475 feet). The limestone was formed from the shells of countless billions of marine organisms, laid down many millions of years ago under the sea, and later uplifted by movements of the earth's crust. Erosion and further movements have produced a fantastic array of caves and other features (described in the section entitled *Climate, Geography and Geology* and illustrated below). Elsewhere are smaller hills and fertile alluvial lowlands. Parts of the park are used by some families of the wandering Penan people, who are able to survive by harvesting wild sago palms for their starch and by hunting wild animals (*see* the section entitled *The Peoples of Malaysia*). The park is open to visitors, by way of an exciting trip up the Baram and Tutoh Rivers.

OPPOSITE PAGE An aerial view of the Gunung Mulu massif.

BELOW A view towards Gunung Mulu from the Tutoh River.

ABOVE A view towards the entrance of the Great Deer Cave, the cave which is nearest to the entrance of Gunung Mulu National Park, and which contains probably well over a million bats. The cave, two kilometres (1¼ miles) long and a minimum of ninety metres (295 feet) high and wide at all points, passes right through the southern range of limestone hills in the park.

FAR LEFT The limestone pinnacles on Gunung Api in Gunung Mulu National Park. Situated in a shallow valley at 1,200 metres (3,900 feet) above sea level, these smooth, razor-edged structures reach forty-five metres (148 feet) in height and are up to twenty metres (65 feet) wide at the base. Separated by deep fissures and low, bushy mountain forest, this curious natural spectacle presents an eery sight in the mists which often envelop the mountain.

LEFT The interior of the Clearwater Cave. A tributary of the Melinau River, named Clearwater, passes through the Api massif, forming a thirty-seven kilometre (23-mile) long passage.

ABOVE Entrance to the Great Deer Cave.

RIGHT Moss-covered stalactites at the entrance to the Clearwater Cave.

ABOVE An unusual lizard, *Gonyocephalus denzeri*, confined to parts of Borneo, and seen here inside the Great Deer cave.

ABOVE RIGHT A tree-gecko (*Cyrtodactylus* species), often to be seen on buttresses or decaying timber.

ABOVE A tiny Bornean toad, *Pelophryne api*, on the leaf of a low forest plant. The species breeds in small puddles of water in hill dipterocarp forest.

ABOVE RIGHT A skink, *Mabuya multifasciata*, often glimpsed on open forest trails.

ABOVE The Common Bluebottle, *Graphium sarpedon* (family Papilionidae). In common with many other butterflies, large numbers of males of these swift-flying 'kite swallowtails' often gather on damp sand or mud, especially if it has been contaminated with urine.

BELOW Mating damselflies of the genus *Neurobasis*. Female damselflies are attracted to the males by their brightly-coloured wings. The wings of the females are unmarked. Damselflies belong to the same order (*Odonata*) as dragonflies.

LEFT Probably a *Trametes* species, this polypore is one of a group of widespread fungi rotting dead, standing or fallen forest trees.

RIGHT The unripe fruits of a fig plant, in this species borne on the main trunk and branches. This feature is known as cauliflory, and is often a characteristic of fruits which are dispersed by bats.

RIGHT The unripe, cauliflorous fruits of a small forest tree, *Baccaurea* species (family Euphorbiaceae). The watery flesh around the seeds of these fruits is very sour, and often used by travellers as a relish for wild meats and fish.

FAR RIGHT The inflorescence of a large forest aroid, *Amorphophallus* species. The plant has an underground tuber from which the single leaf arises, followed by the inflorescence. The flowers are minute and borne at the base of the white spadix within an encircling bract or spathe. The inflorescences of most species of *Amorphophallus* smell disgustingly of carrion, the odour attracting flies which act as pollinators.

The undergrowth herb, *Forrestia mollissima* which belongs to the *Tradescantia* family (Commelinaceae), has dense clusters of white flowers borne along the creeping stem. These are followed by the brilliant purplish fruit shown here. Within the fruits are scarlet seeds, the striking colour contrast probably serving to attract birds of the forest floor which may be responsible for dispersal.

Costus speciosus, a member of the Costaceae, is closely related to the true gingers (Zingiberaceae). It is a coarse herb occurring naturally in light gaps and on river banks and has now become a very common plant of roadsides in forested areas. The large white flowers are followed by these brightly coloured fruits.

Fruits of a member of the moon-seed family (Menispermaceae); the outline of the crescent-shaped seed within the fruits can just be made out. All Malaysian members of the family are forest climbers; they are important as the source of several local medicines and as the food plants of fruit-piercing moths (*see* example on page 66).

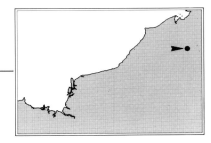

The Kelabit Highlands

The Kelabit Highlands form the headwaters of the Baram River and lie in the true heart of Borneo. Until the advent of the aeroplane, the area was accessible from the coast only by many days of travel by boat and on foot. In the broad valley which lies about 915 metres (3,000 feet) above sea level and which forms the centre of the Highlands are several old settlements, the largest of which is Bario. The Kelabit people who live in this area developed a society and farming methods more complex and stable than those of most of their neighbours further down-river. Traditionally, Kelabits have distinct social classes, although many families lived in a single longhouse without internal partitions. Men of rank wore capes of clouded leopard skin or of bark with hornbill feathers attached. Unlike other native peoples of Malaysia, they grew both hill and wet rice, and were keen on planting vegetables, even before outside influences arrived.

Being near to the border with Kalimantan, they have historical and social ties with their Indonesian neighbours as well as with other Sarawakians. Sarawak natives are renowned for long-distance forest travel, and Kelabits particularly so. Even in today's comparatively easy times, a Kelabit's 'short walk' may be a twelve-hour hard trek for others.

BELOW The forest trail between Bario and Long Dano.

A Shorthorn Grasshopper, probably of the genus *Erucius* (family Eumastacidae).

An adult of the predatory assassin bug, *Eulyes amaena* (family Reduviidae).

A cricket, *Nisitrus vittatus* (family Gryllidae).

A dragonfly, *Orthetrum testaceum* (family Libellulidae).

A harlequin butterfly, *Paralaxita telesia* (family Riodinidae). The forewings of the upperside of this exquisite insect are tipped with a band of carmine red. Found both in Borneo and Peninsular Malaysia, nothing is yet known of the early stages of this species or of its close relatives. The butterflies are found in upland forests (to about 1,220 metres/4,000 feet), where they often settle on leaves in dappled sunlight.

The Glorious Begum, *Agatasa calydonia* (family Nymphalidae, sub-family Charaxinae), patterned in red. The Blue Nawab, *Polyura schreiber* (family Nymphalidae, sub-family Charaxinae) coloured grey and brown. Nawab, a *Polyura* species (family Nymphalidae, sub-family Charaxinae), *above right*. Sergeant, an *Athyma* species (family Nymphalidae, sub-family Limenitinae), *above left*. These four beautiful male nymphalid butterflies are all feeding upon dung on the forest floor. The Glorious Begum is a rare and spectacular insect, found both in Borneo and Peninsular Malaysia.

Pitcher of *Nepenthes stenophylla*, one of the most widespread of the mountain pitcher plants of Borneo.

The elegant pitchers of another interesting mountain species, *Nepenthes reinwardtiana*.

There are many species of *Saurauia* in Borneo. This genus of small trees is related to the chinese gooseberry or kiwi fruit. The delicate flowers are often produced directly from the trunk, rather than from leafy twigs.

The zig-zag flower spike of the epiphytic orchid *Pholidota gibbosa*, a species found throughout tropical South-east Asia.

LEFT A rattan, *Plectocomia mulleri*, often characteristic of forest gaps on infertile soils. The stems are soft and have little commercial value.

BELOW The flower of a *Torenia* species, a delicate herb of wet places.

The flowers of *Rhododendron jasminiflorum*, a lovely species of mid-mountain forest in Borneo, have a spicy sweet jasmine scent.

Burmannia disticha (Burmanniaceae) is a herb of poor heathy soils. This species has green leaves while several members of the genus lack chlorophyll and behave as saprophytes.

Melastoma species, one of the so-called Straits Rhododendrons, not related to the true rhododendrons. There are several species, often occurring as pioneers.

FOCUS ON
SABAH

———◆———

Visitors to Sabah are told that they have come to 'the land below the wind', an expression which is explained by the fact that this state lies just south of a region affected by typhoon winds. Sabah is perhaps equally well characterized as a land of hill ranges and mountains, of which the most outstanding is Mount Kinabalu, at 4,101 metres (13,455 feet) the tallest mountain in South-east Asia. An ancient name, with origins that remain obscure, Kinabalu was adopted to create Kota Kinabalu, Sabah's capital, formerly Jesselton in pre-independence times. And the haunting shape of the mountain inspired the architect of the Sabah Museum to design a unique building. Kinabalu Park, which includes the mountain itself and much forested land to the north, was established in 1964 and covers an area of 75,370 hectares (186,240 acres). Now, nearly 200,000 visitors come to the park every year. Starting from the park headquarters at 1,615 metres (5,300 feet), nearly one-sixth of them make the attempt to reach the summit of Mount Kinabalu on a carefully maintained trail. An annual 'climbathon' was started in 1988, with prizes for those who can reach the summit and return in the shortest time. The fittest take three hours or so. Most visitors take a more relaxed two days, staying overnight in accommodation at Laban Rata or Sayat-Sayat. An even longer trip is better for nature-lovers. The first section of the climb is up a 'staircase' of gnarled tree roots and mossy, orchid-draped boughs. Rhododendrons and pitcher plants soon start to appear. At 2,740 metres (9,000 feet) (Layang-Layang), the soil and flora change rapidly, with distinctive white-flowered *Leptospermum* and the conifer *Dacrydium*. Higher, at 3,200 metres (10,500 feet), the forest becomes thicker and mossy again, with great granite boulders everywhere. After 3,350 metres (11,000 feet), there are very few trees, and the vegetation varies locally with conditions of moisture and shelter. Visitors who do not wish to make the final ascent can see much of interest around the park headquarters, either from the road, or along the eleven kilometres of trails, or in the mountain garden where many local species are brought together for easy viewing and comparison. It is preferable to be accompanied by one of the knowledgeable park staff, otherwise many fascinating aspects of the Kinabalu flora may be missed.

On the eastern flank of Kinabalu Park at about 550 metres (1,800 feet) is the alternative access point at Poring, best known for its hot water spring where a relaxing bath may be taken. Around Poring grows the strange *Rafflesia* plant, a parasite on a wild vine, which has no stem or leaves of its own, and which is only noticed when buds appear from the root or stem of its host. The buds swell over several months to become the size and shape of a compact cabbage before bursting into flower. A walkway in the treetops has recently been constructed at Poring, starting from ground level on a steep slope, and extending horizontally out into the forest canopy. This facility, simple though it is in concept, adds a whole new dimension to enjoyment of the rainforest. Epiphytic plants, flowers and insects can be seen close-to, in their natural setting. Squirrels and birds can be

viewed from the side or even from above, a marvellous improvement over observation from the ground. A route is under construction to link Poring with the main summit trail of Mount Kinabalu; when open, this should prove to be one of the most exciting hiking trails in Malaysia.

Kinabalu Park is proof that rainforests are still, potentially, the source of all manner of valuable natural products. Widespread interest in recent years in wild tropical orchids has revealed the fortunes that can be made by unscrupulous traders, who buy orchids collected from the wild, and export them illegally to buyers in Europe, North America, Australia and elsewhere in Asia. Sabah, including Mount Kinabalu, has suffered badly from this trade. Dedicated people in several countries are doing their best to stem the illegal trade before whole orchid populations, or even species become extinct. They are helped by the Convention on International Trade in Endangered Species of Fauna and Flora (CITES), to which most relevant countries are signatories. Many people believe, however, that it makes more sense to grow rare wild orchids on a large scale in nursery conditions, thus demonstrating to governments the value of rainforest plants other than timber trees, while lowering both the orchids' individual prices and the incentive to plunder wild populations. Not only orchids have a money value. Two species of trees which occur in Kinabalu Park serve as examples. One is a native wild cinnamon (*Cinnamomum burmanii*), known as *kendingau*, and the origin of the name of the town Keningau in Sabah. The bark is good enough to be marketable as cinnamon spice. The Assistant Park Warden at Poring has suggested that this tree be brought into cultivation as a smallholder's crop in the hilly region of Sabah. What an excellent idea, which could help to save a plant species now dying out as a result of forest clearance, at the same time providing income for rural people. The other tree is a lemon (*Citrus halimii*), known only from a few localities in Peninsular Malaysia, and two sites in Sabah. The fruits of the Peninsula trees are rather small but those of the Kinabalu form are equal in size and flavour to the cultivated European lemon, a species that requires a seasonally dry climate. How about bringing this Malaysian lemon into cultivation?

All Sabah's west-coast communities – of Dusun and Kadazan peoples and their rice fields – lie in the shadow of the Crocker Range, the mountain range which includes Mount Kinabalu and which stretches from the northern end of the state to the Sarawak border. At one point, called the Padas Gorge, the Crocker Range is cut right through by the mighty Padas River, which drains the south-western interior part of Sabah and dumps silt into the South China Sea, forming the Klias Peninsula.

OPPOSITE PAGE The Slow Loris (*Nycticebus coucang*), a soft-furred nocturnal primate of the forests and rural areas of Malaysia.

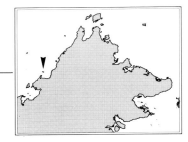

Pulau Tiga Park

Situated in Kimanis Bay off the west coast of Sabah, Pulau Tiga Park was established in 1978 and consists of 15,257 hectares (37,700 acres) of sea and underwater habitat along with three islands: Tiga, Kalampunian Besar and Kalampunian Damit. Pulau Tiga means Three Island, a reference not to the number of islands, but to the shape of the largest one, which consists of three low hills. This distinctive shape reflects the origin of Pulau Tiga, three 'mud volcanoes', which are not true volcanoes, but which originate as mineral-rich mud expelled from deep underground. In 1941, one of them erupted and smothered nearly thirty hectares (70 acres) of the island's forest in boiling mud, with an explosion which was heard over a hundred kilometres away. The 'volcanoes' have since been passive, although small amounts of mud and gas continue to be emitted from the top of one of them. Pulau Tiga is covered in attractive forest, which can be viewed from clear trails. The white sand beaches which fringe most of the island are made up of small coral fragments. The surrounding sea is clear and unpolluted, and nearby coral reefs are barely disturbed. All three islands in the park have remained in pristine condition because there have never been settlements here, although the Sabah Parks authorities and the Sabah branch of the National University of Malaysia now have stations on Pulau Tiga. Pulau Tiga is an important sanctuary for the curious megapode bird (*see* page 164), while Kalampunian Damit is famous in Sabah as Snake Island (*see* illustrations below).

OPPOSITE PAGE A fine sandy beach on Pulau Tiga.

BELOW Sunset from Pulau Tiga.

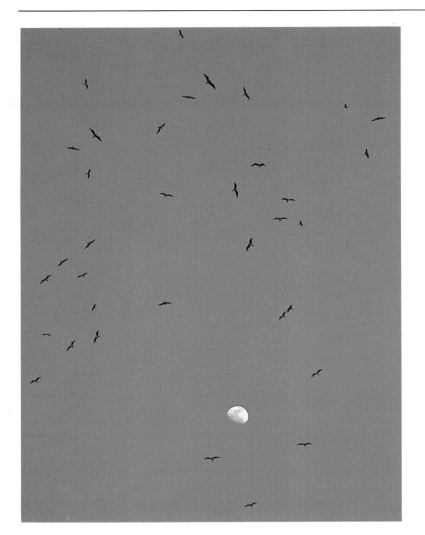

LEFT Lesser Frigate Birds (*Fregata ariel*) over Pulau Tiga. These magnificent fish-eating sea birds move great distances around the coast of Borneo. They roost seasonally on Pulau Kalampunian Damit, near Pulau Tiga.

BELOW A view of Pulau Kalampunian Damit – often known as Snake Island – off Pulau Tiga.

A Yellow-ringed Cat Snake (*Boiga dendrophila*), a species typically of mangrove areas, on Pulau Tiga.

A cluster of Seasnakes (*Laticauda colubrina*) amongst the rocks of Pulau Kalampunian Damit, where they lay their eggs.

A Seasnake (*L. colubrina*) eating an eel on Pulau Kalampunian Damit.

A Grey-tailed Racer Snake (*Gonyosoma oxycephalum*) on Pulau Tiga.

A Water Monitor Lizard (*Varanus salvator*), the main natural predator of the eggs of the rare megapode bird on Pulau Tiga.

A Hermit Crab (family Diogeniadae) on Pulau Tiga.

ABOVE The flower of *Hibiscus tiliaceus*, a small tree of tropical shores world-wide. The tree itself was once an important source of various materials for tropical island-dwellers, including rope and boat caulking (from the fibrous bark), timber, floats, firewood, medicines and food (the young leaves).

ABOVE RIGHT The flower of *Barringtonia asiatica*, a common and attractive seashore tree throughout South-east Asia.

RIGHT The flowers and leaves of an *Ardisia* species, a small, bushy tree on Pulau Tiga.

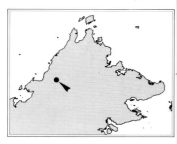

Crocker Range and Padas Gorge

The western side of Sabah is dominated by a range of densely-forested mountains – the Crocker Range – which separates the coastal plains from the remainder of the state. Until the turn of this century, the only way to cross over the range was by walking along one of the several trails maintained by the Dusun and Murut peoples of the interior. The Crocker Range is bisected by a mighty river, the Padas, which drains much of western interior Sabah and which flows through a gorge between Tenom and Beaufort. Boulder-strewn and swift, the waters of Padas Gorge do not permit passage by boats (although nowadays, with safety precautions, parts are navigated for the sheer fun of it). The early British Chartered Company administration of Sabah, in typical enterprising style, decided that the interior of Sabah must be opened up for exploitation and that the way to do so was to build a railway along the side of Padas Gorge. This was completed in 1905 and continues to be used to this day. Now, too, there are four roads over the Crocker Range, from the coastal plains to the towns of Ranau, Tambunan, Keningau and Tenom. The maintenance of the Crocker Range under its natural forest cover is vital to protect clean, constant water supplies for communities in the foothills and plains to both the west and east. Formerly a forest reserve, 139,919 hectares (345,740 acres) of the Crocker Range between Ranau and Padas Gorge was established as a national park in 1984.

BELOW A view eastwards from the Crocker Range, showing the Tambunan Plain and the Trus Madi Range in the distance.

ABOVE A view westwards from the Crocker Range, showing deforested slopes on the foothills.

LEFT The flowers and buds of *Dillenia suffruticosa*, a woody shrub which often colonizes infertile, deforested soils.

OPPOSITE PAGE The Padas Gorge, which bisects the Crocker Range between Beaufort and Tenom. The railway line is just visible on the left.

BELOW The flower of *Rafflesia pricei*, one of three *Rafflesia* species found in the mountain ranges of Sabah. All are parasitic on a particular wild vine, *Tetrastigma*. This *Rafflesia* is restricted to the upper altitudinal range of the dipterocarp forest.

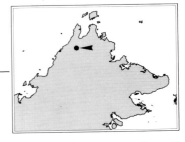

Mount Kinabalu

The grandeur of Mount Kinabalu has inspired the adventurous to climb to its peak since 1851, when Mr Hugh Low mounted the first expedition with a party of forty-two porters. Mount Kinabalu is superlative in many ways. Not only is it the highest mountain in South-east Asia, it also has an outstanding array of flora and fauna. After thirty-five years of experience as a renowned tropical botanist and after joining a major botanical expedition on the mountain in 1961, Professor E. J. H. Corner declared that he believed the mountain to have 'the richest and most remarkable assemblage of plants in the world'. Kinabalu certainly has the greatest known concentration of wild orchid species, probably about 1,200 of them. There are also more members of the magnolia family here than in any other comparable area. Rhododendrons, pitcher plants, figs, mosses and ferns occur in unusual diversity and abundance too. Many curious remnants of past, colder climates in the tropical region are found on the mountain: buttercups, violets, various other herbaceous plants, even a species of maple tree, all can be seen on Kinabalu. There are some species of plants and small animals which are known only from here and nowhere else. In 1964, Mount Kinabalu was gazetted as a national park, incorporating Mount Tambuyukon to the north and associated foothills.

OPPOSITE PAGE The upper part of Mount Kinabalu, showing the twin peaks known as the 'Donkey's Ears' (highest point 4,054 metres/13,301 feet).

BELOW At 4,101 metres tall (13,455 feet), and still rising imperceptibly, Mount Kinabalu is the highest peak between the Himalayas and the island of New Guinea. It dominates the landscape of Sabah, and is perhaps the grandest physical feature of Malaysia.

PREVIOUS PAGES A view over the Crocker Range from Mount Kinabalu (at altitude 3,200 metres/10,500 feet).

OPPOSITE PAGE A view towards Tunku Abdul Rahman Peak (3,948 metres/12,952 feet) from the stunted vegetation at 3,200 metres (10,500 feet).

RIGHT The clear water of a forest stream at about 1,520 metres (5,000 feet) on Mount Kinabalu.

BELOW The forest at about 2,750 metres (9,000 feet) on Mount Kinabalu, draped with mosses and other epiphytes. The dead trees probably perished as a result of drought.

BELOW RIGHT The Mountain Treeshrew (*Tupaia montana*), a species restricted to the mountain ranges of north-western Borneo. It feeds on insects and fruits.

LEFT A view in the montane forest at about 1,580 metres (5,200 feet) on Mount Kinabalu. Visible are an epiphytic fig (*Ficus* species), a pandan and a *Coelogyne* orchid.

BELOW LEFT A wild hybrid rhododendron (between *Rhododendron buxifolium* and *Rhododendron rugosum*), at about 3,500 metres (10,000 feet) altitude. The hybrid – a bushy tree – is commoner than either parent plant on the exposed areas around Panar Laban.

BELOW *Rhododendron crassifolium*, an epiphytic species which thrives in forest around the Kinabalu Park headquarters. The flowers vary in colour from red to pale pink or even white.

A shrub, *Medinella* species (family Melastomataceae), near the park headquarters on Mount Kinabalu.

The Heath Rhododendron (*Rhododendron ericoides*), a distinctive species which grows only at the highest levels of Mount Kinabalu, in cracks between boulders.

Rhododendron fallacinum, with deep orange-pink flowers that bear numerous tiny scales. This plant grows above 1,800 metres (6,000 feet) on Mount Kinabalu.

Phyllagathis species (family Melastomataceae), at 1,580 metres (5,200 feet). There are at least twelve species of this genus on Mount Kinabalu.

LEFT An orchid, *Dendrochilum* species, with rows of delicate small flowers, at 2,750 metres (9,000 feet) on Mount Kinabalu.

ABOVE Mountain Schima (*Schima brevifolia*), a small bush at 3,350 metres (11,000 feet) on Mount Kinabalu. The young leaves are reddish brown.

BELOW The Maiden's Veil Fungus (*Dictyophora duplicata*) at 1,580 metres (5,200 feet) on Mount Kinabalu. This bizarre fungus relies on insects to disperse its spores. The insects feed on the vile smelling mucous on the outer surface of the cap, their feet becoming contaminated with spores which are then transported elsewhere.

BELOW The splendid pitchers of *Nepenthes villosa*, clearly displaying the very coarse ribs round the pitcher mouths. This species occurs in some abundance at about 2,750–3,000 metres (9,000–10,000 feet) on Mount Kinabalu, higher than any other species.

RIGHT *Schefflera* species, a member of the ivy family, said to have medicinal properties, at 1,580 metres (5,200 feet) on Mount Kinabalu.

BELOW Kinabalu Balsam (*Impatiens* species) at 1,800 metres (6,000 feet) on the mountain.

ABOVE A small undergrowth palm, *Pinanga capitata*, at 1,580 metres (5,200 feet) on Mount Kinabalu.

RIGHT A small woody shrub, *Vaccinium stapfianum*, at 3,350 metres (11,000 feet) on Mount Kinabalu.

ABOVE A Small-clawed Otter (*Amblonyx cinerea*) eating fish. This species of otter is adaptable, feeding on all manner of freshwater life. It occurs along streams in Sepilok Forest Reserve.

BELOW The Moonrat (*Echinosorex gymnurus*), a nocturnal mammal which feeds largely on earthworms and other invertebrates. Its almost-total white coloration is unique amongst Malaysian mammal species.

BELOW The Green Imperial Pigeon (*Ducula aenea*), one of the largest Malaysian pigeons, a resident of extensive lowland forests. It has a distinctive call, reminiscent of a distant, high-pitched mooing of cows.

RIGHT A Pig-tailed Macaque (*Macaca nemestrina*), one of several monkey species in Sepilok Forest Reserve. It is most common in hill dipterocarp forest, but is adaptable and often raids plantations and gardens.

BELOW The Brown Leech (*Haemadipsa zeylanica*) after a meal of blood, its sole food. This creature, distantly related to earthworms, lives on the ground in Malaysian dipterocarp forests. It seeks ground-dwelling birds and mammals, including humans, to supply its food. Its bite is painless, as it injects anaesthetic as well as anticoagulant.

BELOW RIGHT The Painted Leech (*Haemadipsa picta*), a colourful green, orange and black species which sits on the leaves of low vegetation and awaits victims to brush past. Its bite produces a sharp stinging sensation.

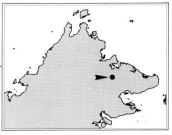

The Kinabatangan River

At 560 kilometres (348 miles) long, the Kinabatangan is Sabah's longest river, having its sources in the steep mountain ranges of the south-west and flowing out into the Sulu Sea through the largest tract of mangrove in Malaysia. The banks of the Kinabatangan are sparsely inhabited and this seems always to have been so. Nevertheless, for centuries the river has been a highway for the transportation of forest produce from its hinterland down to trading boats in the Sulu Sea. Around 1812, an Englishman named John Hunt estimated that an astounding 37,000 kilograms (81,500 pounds) of wild beeswax (used mainly to make candles) and 23,000 kilograms (50,700 pounds) of edible bird's nests came annually from the Kinabatangan and Sandakan region. The lower Kinabatangan region does indeed contain several caves with edible birds' nests, including Gomantong, the most important nesting site in Malaysia. Today, timber is the only forest produce which comes down the river in large quantities, in the form of tree trunks lashed together as enormous, floating, herring-bone pattern rafts. Most of this timber has come from forests on dry land away from the main river, and that land which is not too steep is now being converted to plantations of cocoa and oil palm. Low-lying land is per-

manently waterlogged, however, and covered in freshwater swamp forest, a habitat poor in commercial timber trees, dotted with natural lakes. Coupled with irregular floods, this has meant that much of the region remains under natural forest. And this forest contains the largest Malaysian populations of elephants, Orang-utans, Proboscis Monkeys and snakebirds, as well as much other wildlife.

BELOW Forest near the lower Kinabatangan River, a scene typical of many lowland areas in eastern Sabah. Much of the forest shown here was burned by fires which swept through parts of Sabah during an unusually long drought in 1983. Since then, the forest has regrown. A tall *tualang* tree may be seen on the right-hand side.

Elephants (*Elephas maximus*) on a rural road in eastern Sabah. The largest concentrations in Malaysia occur in the lower Kinabatangan region, where herds of forty or more are commonly seen.

ABOVE An Estuarine Crocodile (*Crocodylus porosus*), the largest reptile in Sabah, which occurs in the Kinabatangan and other large rivers. It is scarce, shy and rarely seen. Nests are made in secluded spots by the female, in which she lays and guards her eggs.

FOLLOWING PAGES The Segaliud River, which forms a boundary between the lower Kinabatangan and Sandakan districts. The tall trees in this picture are red durians (*Durio graveolens*), a wild species which is planted in the old villages of the area.

ABOVE The Bornean Gibbon (*Hylobates muelleri*) in the forest of Danum Valley. This area contains all the primate species known to occur in northern Borneo except the Silvered Leaf Monkey.

LEFT A Common Treeshrew (*Tupaia glis*), one of five treeshrew species in the lowland forests of Sabah, here feeding on a grasshopper. These creatures resemble the earliest small mammals which evolved and spread world-wide many tens of millions of years ago. They are now restricted to South-east Asia.

ABOVE The Clouded Leopard (*Neofelis nebulosa*), largest wild cat of Borneo. Rarely seen but not uncommon in extensive forest areas, this cat preys on mammals ranging in size from rats to wild pigs.

RIGHT The Leopard Cat (*Felis bengalensis*), a wild cat seen most often on forest fringes and logging roads in eastern Sabah.

RIGHT The Greater Mouse-deer (*Tragulus napu*), one of two species of mouse-deer which occur in the Malaysian dipterocarp forests. They are common, but shy and cryptic except at night, when their presence is betrayed by their eyeshine in the beam of a lamp.

LEFT A Malay Weasel (*Mustela nudipes*), one of the smallest carnivores of the dipterocarp forests. This species spends much of its time in holes under the ground, emerging periodically during the daytime to seek its prey.

MIDDLE LEFT The Common Palm Civet (*Paradoxurus hermaphroditus*), the commonest and most adaptable of nine species of nocturnal civets which occur in Sabah. This species is active both in trees and on the ground, seeking fruits and small animals as food.

BOTTOM LEFT A Pangolin (*Manis javanica*) on the forest floor. This strange, scaly mammal with no teeth feeds on termites and ants, both in trees and on the ground. When disturbed, it rolls itself into an impenetrable ball.

ABOVE A tree frog (*Polypedates otilophus*) at Danum Valley.

LEFT The Southern Pied Hornbill (*Anthracoceros convexus*), a handsome, noisy bird which travels in small flocks through the forest in coastal areas and along large rivers in Malaysia. It feeds on fruits and invertebrate animals.

ABOVE The fruits of a climbing fig (*Ficus* species). Unlike the more common and massive strangling figs, this variety grows in the form of single stems which grow up tree trunks from the ground. The large fruits look attractive but are inedible to humans, being hard and containing a sticky latex.

BELOW The flowers of a small, woody legume plant at Danum Valley.

ABOVE Pill millipedes (order Oniscomorpha), shiny, hard invertebrates which live on the forest floor amongst leaf litter, foraging on various vegetable matter. They roll into a ball to protect themselves from predators.

ABOVE RIGHT A flat-backed millipede of the genus *Platyrhachus* (family Platyrhachidae). Members of this group are often found under the bark of rotting trees in forest floor litter. They are thought to feed on fungi and vegetation detritus, but not on the wood itself. They are medium-sized, usually 5–10 centimetres (2–4 inches) in length.

ABOVE A scrambling plant, *Merremia* species, which colonizes landslips and logging roads, helping to check soil erosion. If it grows too densely, regeneration of forest is inhibited, and it is sometimes a serious problem in managing the natural regrowth of logged forests.

RIGHT A cup fungus, probably *Cookeina tricholoma*, in lowland dipterocarp forest.

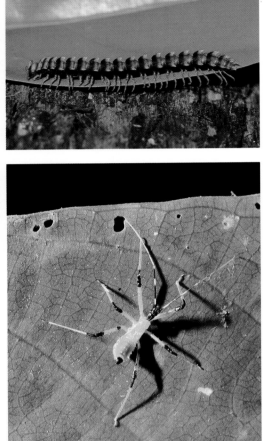

ABOVE The nymph of a predatory assassin bug belonging to the genus *Velinus* (family Reduviidae). Almost all members of the Reduviidae, a very large bug family indeed, are predatory upon other small arthropods, piercing their prey's exoskeleton with the recurved and stout rostrum and ingesting liquid from the body of the prey.

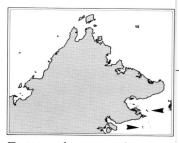

Semporna and Sipadan Islands

Dotting the sea to the east of the Semporna Peninsula in south-eastern Sabah is a long scattering of islands, most of which belong to the Philippines. Those within forty kilometres (25 miles) of the Sabah mainland belong to Malaysia, and are known as the Semporna Islands. All lie on the continental shelf and most consist of old volcanic rocks, some capped with limestone. About thirty-two kilometres (20 miles) to the south of the Semporna Peninsular is a single island, off the continental shelf, called Sipadan. It is the only truly oceanic island in Malaysia, rising as a single outcrop from a depth of 610 metres (2,000 feet). The Semporna Islands and surrounding waters are amongst the most beautiful in Asia, with a rich diversity of marine life. A complex mix of Bajau and Suluk peoples inhabit the region, most with cultural and historical links closer to the present-day Philippines than to Malaysia. Unfortunately, their presence makes con-serving the area an equally complex problem. Coral reefs throughout the world have suffered greatly from human activities – overfishing, disturbance from tourists, bombing, pollution and siltation, as well as removal for landfill and road stone. Sipadan is rated as one of the least disturbed – possibly *the* least disturbed – of coral reefs in the world. Sharks and turtles abound in the nearby waters, and the island's sandy shore is an important nesting ground for Green Turtles.

BELOW One of the Semporna islands, where coral reefs form extensive beds under shallow blue water.

ABOVE LEFT Sweetlips or Rubberlips, *Plectorhinchus* species. These fishes, formerly called *Gaterin*, are found only in the Indo-West Pacific. They are abundant on reefs, usually around the edges, and in sandy bays. There are numerous species, many of which have conspicuous oblique rows of spots on a light (often yellow) background.

ABOVE A Hawksbill Turtle (*Eretmochelys imbricata*), one of the two species of marine turtle which frequent the Semporna Islands. The Hawksbill, smaller and less common, feeds on invertebrate animals of the coral reefs.

LEFT A sponge on the corals of Sipadan Island, one of the many forms of life which make up the coral-based community, taking advantage of the constant, nutrient-rich environment.

BELOW LEFT Sea anemone and the anemone fish *Amphiprion* (probably *Amphiprion frenatus*). These fishes are immune to the stinging cells of the anemone because their body slime lacks the protein that triggers the sting. They live in pairs close to and within the anemone, breeding and laying their eggs on the rock beside it.

BELOW A starfish on coral off Sipadan Island.

ABOVE Damsel fish and fusiliers off Sipadan Island. The damsel fish include *Dascyllus trimaculatus* and *Anthias* species and the fusiliers are species of *Pterocaesio*. All are common on coral reefs.

BELOW A fan coral, one of several hundreds of coral species known from Sabah waters.

RIGHT Bodgaya Island, the remains of an extinct volcano off Semporna, and a school of flying fish.

The orchid *Vanda dearei* has sweetly fragrant flowers.

Phalaenopsis pantherina, a relative of the commonly cultivated moth orchid, *P. amabilis*.

The flowers of *Vanda scandens* are strongly fragrant.

Dyakia hendersoniana has a neat flower spike.

Like many Bornean epiphytic orchids related to the genus *Sarcanthus*, this *Pteroceras* has small flowers of great beauty.

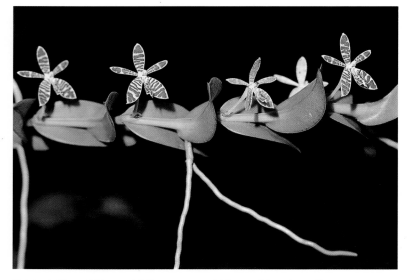

Trichoglottis smithii, a wild Bornean orchid, would make a fine ornamental.

There are estimated to be over two thousand species of wild orchid on the island of Borneo. About one half of these occur on or near Mount Kinabalu. It is not surprising, therefore, to find that Sabah and Sarawak, especially the former, attract as much interest from orchid lovers as any place in the world. The rarity of some of the Borneo species, coupled with the increasingly popular interest in them, means that they fetch high prices, especially in Europe and North America. This has caused the plundering of orchids from many forests. Amongst the most sought-after is *Paphiopedilum rothschildianum*, now known to occur for certain only at localized sites around Mount Kinabalu. More widespread, but equally prized, are the *Dimorphorchis* species, which bear two forms of flower on one long stem. In contrast, many orchids, such as *Dendrochilum* and *Pholidota* species, have tiny, delicate flowers well under a centimetre wide.

This *Dendrobium* species belonging to the section *Eugenanthe* produces its flowers from bare stems.

The recently discovered and aptly named *Renanthera bella* is found only where the forest soil is rich in metals such as iron and manganese.

One of the finest of the moth orchids, *Phalaenopsis violacea*, much used by orchid hybridizers, and often over-collected in the wild.

Basal flower of the orchid *Dimorphorchis lowii*, strikingly different from the upper flower (*see* page 118, *top right*).

Conservation Areas in Malaysia

The only way in which most wild species of plants and animals can be conserved in perpetuity is to protect by law an array of all their natural habitats. The following lists of conservation areas include most of the National Parks (State Parks in Sabah), Wildlife Reserves and Sanctuaries established in Malaysia by early 1990. Some 'conservation areas' are not included because they serve more for recreation, research or education than for conservation (for example, Templer Park and Pasoh Reserve in Peninsular Malaysia and Semengoh Wildlife Rehabilitation Centre in Sarawak). Many vital areas are left out because they are not established primarily for species conservation and are too many to list in full. Into this category come most of the Forest Reserves for timber production and water catchment protection, throughout Malaysia. However, the lists provided below do include a sample of conservation areas other than the 'official' parks and wildlife areas. Some show how habitats modified by human use can be important for wildlife. Examples are Kuala Selangor Nature Park and Matang Forest Reserve on Peninsular Malaysia's west coast and the Gomantong and Madai-Baturong Forest Reserves in eastern Sabah, which protect the caves used by edible birds' nest swiftlets. Some areas, like Maligan, Tavai and Silam, also in Sabah, are retained under natural forest primarily to protect water supplies, but at the same time they preserve a wealth of unique plant species. In addition, some proposed conservation areas in which the relevant governments have already shown strong interest are named for Peninsular Malaysia and Sabah. More than ten National Parks, Wildlife Reserves and Sanctuaries have been proposed for Sarawak, but details are not available at present.

PENINSULAR MALAYSIA

Existing	Size (in hectares)
Ampang Forest Reserve	2,700
Cameron Highlands Wildlife Sanctuary	64,953
Fraser's Hill (Bukit Fraser) Wildlife Sanctuary	2,980
Kerau Wildlife Reserve	53,095
Kuala Selangor Nature Park	260
Kuala Selangor Wildlife Reserve	44
Matang Forest Reserve	40,711
Pahang Tua Bird Sanctuary	1,336
Pulau Tioman (Tioman Island) Wildlife Reserve	7,160
Sungai Dusun Wildlife Reserve	4,330
Sungkai Wildlife Reserve	2,426
Taman Negara (National Park)	434,350

Proposed
Endau-Rompin State Parks (Johor and Pahang)

In addition, the maritime waters within eight kilometres of Redang Island and within three kilometres of the following offshore islands are at present legally protected from exploitation, and under consideration as Marine Parks: (Kedah) Lembu, Kaca, Payah and Segantang; (Johor) Babi Besar, Babi Hujung, Babi Tengah, Rawa, Tinggi, Mentinggi and Sibu; (Pahang) Tulai, Cebeh, Sembilang, Seri Buat and Tioman; (Terengganu) Perhentian Kecil, Perhentian Besar, Kapas, Lang Tengah and Tenggol.

SARAWAK

Existing	Size (in hectares)
Bako National Park	2,728
Gunung Gading National Park	5,430
Gunung Mulu National Park	52,887
Lambir Hills National Park	6,952
Lanjak-Entimau Wildlife Sanctuary	168,755
Niah National Park	3,140
Pulau Tukong Ara-Banun Wildlife Sanctuary	1
Samunsam Wildlife Sanctuary	6,092
Similajau National Park	7,067

SABAH

Existing	Size (in hectares)
Crocker Range National Park	139,919
Danum Valley Conservation Area	42,755
Gomantong Forest Reserve	3,297
Kinabalu Park	75,370
Kota Belud Bird Sanctuary	12,000
Kulamba Wildlife Reserve	20,682
Madai-Baturong Forest Reserves	5,867
Maliau Basin Conservation Area	39,000
Maligan Virgin Jungle Reserve	9,240
Mount Silam	4,128
Pulau Tiga Park (islands and sea)	15,864
Sepilok Forest Reserve	4,294
Tabin Wildlife Reserve	122,530
Tavai Protection Forest Reserve	22,697
Tawau Hills Park	27,972
Tunku Abdul Rahman Park (islands and sea)	4,929
Turtle Islands Park (islands and sea)	1,740

Proposed
Kinabatangan Wildlife Sanctuary
Likas Swamp
Semporna Marine Park (islands and sea)
Sipadan Marine Park (island and sea)

Index